CHEMISTRY
THE IMPURE SCIENCE

CHEMISTRY
THE IMPURE SCIENCE

Bernadette Bensaude-Vincent
Université Paris X, France

Jonathan Simon
Université Lyon 1, France

Imperial College Press

Published by

Imperial College Press
57 Shelton Street
Covent Garden
London WC2H 9HE

Distributed by

World Scientific Publishing Co. Pte. Ltd.
5 Toh Tuck Link, Singapore 596224
USA office: 27 Warren Street, Suite 401-402, Hackensack, NJ 07601
UK office: 57 Shelton Street, Covent Garden, London WC2H 9HE

Library of Congress Cataloging-in-Publication Data
Bensaude-Vincent, Bernadette.
    Chemistry : the impure science / Bernadette Bensaude-Vincent, Jonathan Simon.
        p. cm.
    Includes bibliographical references and index.
    ISBN-13 978-1-84816-225-9
    ISBN-10 1-84816-225-1
    1. Chemistry. 2. Chemistry--Philosophy. I. Simon, Jonathan, 1964–
    QD31.3 .B455 2008
    540--dc22
                                                    2009275331

British Library Cataloguing-in-Publication Data
A catalogue record for this book is available from the British Library.

First published 2008
Reprinted 2010

Typeset by Stallion Press
Email: enquiries@stallionpress.com

Printed in Singapore by World Scientific Printers

To Larry Holmes

In Memoriam

ACKNOWLEDGEMENTS

Both authors would like to start by thanking each other for the many fruitful intellectual exchanges both prior to the present project and during the preparation of *Chemistry, The Impure Science.*

The authors would like to thank Phillippe Pignarre, editor of the series *Les empecheurs de penser en ronde,* and the publisher Seuil, for granting permission to use Bernadette Bensaude-Vincent's book, *Faut-il avoir peur de la chimie?* (Paris, Seuil, 2005), which served as the inspiration for the present work.

Bernadette Bensaude-Vincent would like to acknowledge her debt to Isabelle Stengers for her continuous inspiration and intellectual exigency. She is also grateful to all the participants in her annual seminars in the history of chemistry at the philosophy department of the University of Paris X — Nanterre. The reflections and expertise brought together in these seminars have significantly contributed to the present work. This book has also benefited from her interviews with a number of chemists, in particular Hervé Arribart, Jacques Livage, Michel Pouchard, and George Whitesides as well as collaborations with colleagues, notably William Newman and Arne Hessenbruch, who have shared her interest in impure sciences.

Jonathan Simon gratefully thanks Ted McGuire and Peter Machamer, who "scrubbed his flanks" at the University of Pittsburgh. He also owes a great debt to Fritz Ringer, whose premature death deprived the academic world of a model of intellectual rigour. Nor will he soon forget his fellow students in Pittsburgh, in particular Ofer Gal, Andrea Woody, Silvia Castro, William Sutherland, Rachel Ankeny, Heather Douglas, and Michel Janssen. Although they probably would not want to be reminded of it, John Worrall, John Earman, John Norton, and Clark Glymour also contributed to shaping Jonathan Simon's philosophical perspective on science. Along the long road to secure employment, he has received support and encouragement from Hans-Jörg Rheinberger, Delphine Gardey, Nic Rasmussen, Jean-Paul Gaudillière, Volker Hess, Patricia Caillé, Axel Hüntelmann, John Ceccatti, and Christian Bonah. Last, but not least, a debt is due to Jessie, Owen, Alan, and Patricia for their invaluable assistance in the preparation of the text.

The authors are pleased to include a number of images in the volume that serve as more than incidental illustrations. In assembling these images, we were helped by the staff of the Chemical Heritage Foundation in Philadelphia, Eva Wittig at Evonik Industries, Marta Lourenço at the Museum of Science of the University of Lisbon, Portugal, Michael Green and Frank Norman at the National Institute for Medical Research in England, as well as staff at the *Bibliothèque Nationale Universitaire de Strasbourg* and the *Service Commun de la Documentation* at the Université de Strasbourg. Thanks also to Anne Denuziere at the School of Pharmacy of the Université de Lyon for the information on the analysis of ions. Finally, the authors would also like to thank the editorial team at Imperial College Press for helping to produce a high-quality book in such a short time.

Bernadette Bensaude-Vincent & Jonathan Simon
May 2008.

CONTENTS

CHAPTER 1

INTRODUCTION: CHEMISTRY AND ITS DISCONTENTS

Look around you! How many of the objects in your immediate environment are products of synthetic chemistry? Of course, the answer to this question will depend on where you are. For those of our readers fortunate enough to be trekking through a rain forest in Costa Rica, perhaps the clothes you are wearing and the ink on the pages of this book are the only such products. For the majority of you reading this book in your home, your office or your classroom, much of what surrounds you is made of a combination of synthetic polymers, usually coloured using synthetic dyes made in factories that transform petroleum into the rainbow of colours so characteristic of contemporary interior design. In many modern environments — the interior of an aeroplane, a train or a car — it is probably easier to try to pick out the few materials that are not synthetic chemicals (metal, wood, leather, cotton, wool, brick, plaster, etc.) and assume that the rest is fabricated from petroleum-based rubbers, plastics or other synthetic polymers. Even natural materials are now usually covered by some synthetic coating, and few paints or lacquers are made entirely of naturally occurring materials.

Whether we like it or not, chemistry has transformed our lives: refining the petrol for our cars, providing the microchips for our computers, and producing medicines for chemotherapy, to cite but three examples. It would be hard for someone in the industrialized world to imagine a world without the contributions of this particularly productive science. Nevertheless, this claim that chemistry has transformed our lives — usually accompanied by the implicit message that it has changed our lives for the better — evokes

1

a cynical response from many people. While we have to accept the evidence of our own eyes that chemical products are everywhere, and accept that it is today impossible to close the Pandora's box that is the synthetic chemical industry, there are many who suspect it is not an unmitigated good and that those who argue for the positive image of the industry are not being entirely honest with the public or with themselves.

The most sensitive point of criticism for modern chemistry is undoubtedly the issue of environmental pollution. Indeed, chemistry has been implicated either directly or by association in many of the most prominent cases of industrial pollution. The chemical industry has a long history of polluting the air, water and soil, which dates back several centuries. Considering only the twentieth century, it has been responsible for a number of major catastrophes including the mercury poisoning that killed thousands in Minamata in Japan as well as the escape of deadly gas in Bhopal in India, which we shall be discussing in more detail in Chapter two. Events like these have turned the chemical industry into an emblem for the relationship of exploitation that exists between capitalists and labourers, between rich and poor, and, more recently, between North and South. Of course, many industries exploit inequalities between the West and the developing world (manufacturers of shoes, toys and clothing, to name but three) in order to increase their profits or simply to survive. Nevertheless, due to its long history of environmental pollution and the ubiquity of synthetic waste products such as plastic bags, rubber tyres, etc. in the industrialized world where chemical goods are consumed, chemistry has become a particularly visible symbol of this inequality and exploitation. Chemistry is too often seen as an impure science; contaminating the soil, poisoning our water, and polluting the air that we breathe.

The Philosophy of Chemistry

This image problem is not confined to the chemical industry, however, but also affects the academic discipline of chemistry. The subaltern status of chemistry is as old as the discipline itself. Even as it first established itself as an independent scientific discipline in the eighteenth century, it was regarded as being intellectually inferior to mathematics and physics.[1] Chemistry continues to be perceived as a dirty, messy science that lacks the rigour associated with physics, its disciplinary neighbour, qualifying it once again as an impure science. Twentieth-century physicists were only too ready to repeat Rutherford's alleged condemnation of all other disciplines: "there is only physics, the rest is stamp collecting". In more recent times, chemists have seen physics fall from grace in many universities and research institutes. The end of the cold war put an end to many ambitious physics-based projects like Ronald Reagan's Star Wars initiative. As a result, student enrolment in physics has dropped across many American campuses, mirroring this reduction in funding. No doubt the most significant symbolic setback for modern physics was the refusal of the American Congress to fund the Superconducting Super Collider in 1993. Of course, one does not need to be an astute observer of science to remark that it was not chemistry that took the lead from physics as the 'hot' science at the close of the twentieth century, it was biology, and more specifically genetics. Indeed, chemistry has never assumed great prestige in the context of the university, and the history of this neglect of chemistry, whether merited or not, will form one of the central themes of this book.

Despite the changing academic hierarchies of the modern sciences measured in terms of funding and salaries, for many people theoretical physics remains at the top. This is particularly true for philosophers, whose primary interest in science is to find the keys for solving the "big" philosophical questions that have dominated the history of Western philosophy. What is the ultimate nature of the cosmos? Where do we come from? How does our universe work? etc. From this perspective, chemistry holds little interest for philosophers. A central aim of the current volume is to try and turn this philosophical perspective on its head. We will be making a strong argument for the philosophical interest of chemistry based precisely on the fact that it is an "impure science"; that it mixes science with technological applications, that it eschews high theory, and that it does not hold consistency to be its highest value. Philosophers have all too often denigrated chemistry because they considered its methods and achievements from the standpoint of the standards and values of physics. We want to argue that looking more closely at the chemists' practical approach is more philosophically interesting than applying tired philosophical dogmas associated with an ultimately unfruitful reductive vision of science.

We are far from being the first to adopt such an approach. In this respect, the present book is similiar to other works in the philosophy of chemistry, notably those of Davis Baird, Eric Scerri, Joachim Schummer and Jap van Brakel.[2] Although we do not agree with these authors in every respect, we are nevertheless arguing in the same sense, hoping to establish an independent philosophy of chemistry. Our ultimate aim is to challenge the hegemony of a particular positivist conception of the philosophy of science as it is taught in philosophy departments around the world.

The Image of Chemistry

Why, one might ask, does chemistry suffer from this particular negative image? To add insult to injury, despite its apparently transgressive position, chemistry does not even seem to convey the excitement associated with other sciences. Watching the film *Matrix*, we can see that even computing, which is not generally regarded as the most exciting subject in the modern curriculum, has a thrilling "dark side," mirroring the fact that computers and robots have become emblematic of humanity's potential to overcome and improve its lived environment, rather than destroying it. While in the 1950s it was atomic physics that was at the root of science-fiction thrillers, not since Mary Shelley's *Frankenstein* has chemistry inspired this kind of fearful excitement.

A long history lies behind the particular image we have developed of chemistry. Today, this vision is changing in response to the rise of the nanotechnologies, which have opened up new perspectives in terms of the public's appreciation of chemistry. In the pages that follow, therefore, we will discuss the question of why chemistry suffers from an image problem by examining various philosophical questions with the aid of a number of historical reflections. Thus, by the end of the book, we hope to have put the tools in place for responding to the question of why chemistry has a particularly bad image. This analysis requires a long detour through the history and the philosophy of chemistry, which reflects the fact that the public image of chemistry does not result from a specific and fixed modern situation but has deep cultural roots.

The idea that chemistry is an impure science does not only come from its links to pollution. Chemistry is also considered impure because of its hybrid nature, its constant mix of science and technology. As we will be arguing in what follows, chemistry serves as the archetypal techno-science, unable to restrict itself to the high-ground of pure theory, but always engaged in productive practice. When we look back to past philosophers like Denis Diderot or Gaston Bachelard, we can see that the idea that there are two kinds of science — theoretical and practical — is nothing new. Indeed, Diderot explicitly favoured empirical sciences that relied on the work of the hand over pure theory, condemning the construction of theoretical systems as ultimately fruitless. Nevertheless, in the course of the last two centuries, the rise of modern physics has promoted pure theory over other forms of science, making it natural to characterize those that rest at the level of practice as impure if not degenerate. Of course, considering chemistry as impure is ironic in light of the fact that one of the central goals, if not the major obsession of chemistry, has been to purify substances.

The Goals of the Book

What we do in this book, therefore, is to reflect on the image of chemistry by referring to the philosophy of chemistry. Furthermore, we anchor this philosophy of chemistry in a reading of the history of chemistry. Nevertheless, this is not an introduction to the philosophy of chemistry in the traditional sense of a straightforward textbook that "objectively" summarizes current positions, but rather an introduction to our own particular philosophy of chemistry which stresses the fact that this philosophy emerges from the chemist's constant practical engagement with matter. Nor is it intended to be a philosophically oriented introduction to the history of chemistry, there are already books available for those who are seeking this kind of treatment.[3] The present work is best described as a historically based introduction to the philosophy of chemistry, and, as we have already said, our own particular interpretation of this philosophy. While in the pages that follow, we do raise some of the most important issues relating to chemistry that have marked the history of philosophy, we will not, as should become clear, simply offer new responses to the questions that are traditionally posed. Instead, we aim to elaborate a new approach to the philosophy of chemistry, answering questions when we feel that they are relevant, but otherwise not hesitating to shift the ground of debate elsewhere. This is the means we use to elaborate a new philosophy of chemistry. If, by the end of the book you are convinced, or are prepared to entertain the proposition that you can know the material world as well if not better by making and doing things rather than by constructing theories, then we will have succeeded in conveying our central message.

The Structure of the Book

Following this brief introduction, we explore the negative image of chemistry starting with its reputation as a polluting science. Thus, we consider the pesticides and plastics that were once considered the heralds of a brave new world, but have since come to be regarded as chemistry's harbingers of doom. These synthetic chemicals were originally produced and sold as a panacea, meant to satisfy the booming demands of the post-war consumer. However, in less than fifty years, we have seen them acquire an unenviable reputation as poisons and pollutants. Thus, we set the stage of chemistry as an impure science in the sense of a science wedded to a global industry generating toxic substances that have indelibly marked our planet. In Chapter three, we place this tarnished image in the context of a much longer history. Here, we consider the heritage of the alchemical tradition, and in particular, the unbridled ambitions of those who sought to transcend nature by transforming it. Indeed, in successive parts of this book, we follow the Faustian ambitions of challenging nature, and ultimately, imitating life itself, into nineteenth-century organic chemistry, which, through the idea of total synthesis, revived many of these ancient ambitions in a new form. Finally, we ask whether nanotechnology has inherited the same or similar ambitions.

The next section takes us into the characteristic space of the research chemist: the chemical laboratory. Originally the exclusive realm of the chemists, the laboratory remains their privileged site of practice, a place where they produce both theory and substances. Indeed, we want to place special emphasis on this idea that theory and substance are co-produced by the chemist in the laboratory. In this realm, the chemists' intuition — or their tacit knowledge — is sovereign and leads them to transcendent feats of productivity, not only in the transformation of the material world, but also, as we want to emphasize, in the generation of theory. It is this theory, indissociable from an intimate contact with the potential and limits of the material world, that is at the base of a philosophy characteristic of chemistry. We argue that while this distinctive philosophy has been dismissed by philosophers as an impure product of an impure science, it should instead be seen as the central appeal of chemistry for philosophers. We believe that it is up to philosophy to re-assess and try to understand chemistry rather than it being up to chemists to fit their science into the outworn axiomatic mould of traditional philosophy of science.

We continue this reflection on the importance of the laboratory and experimental culture for the chemist with a more detailed case study from the end of the eighteenth century. Thus, in Chapter five, we consider Lavoisier's spectacular public demonstrations of the analysis and the synthesis of water, and how he marshalled the forces of experiment to convince others of his views of matter. Chapter six leads us into a consideration of organic synthesis as a field where theory and practice are particularly entwined. It is in the context of organic synthesis in particular that reactions are developed as useful tools, with industrial applications never far from chemists' considerations. We see how the creative power of organic chemists even led them to dream of creating the whole of life from scratch.

With Chapter seven, we enter into more familiar philosophical territory. Here, we address the problem of the "mixt" with respect to the various traditions of elements that have traversed the history of chemistry. The mixt — a chemical combination composed of elements but not bearing the same properties as the constitutive elements — is a core issue for any philosophy of matter. We argue for the value of the Aristotelian approach in this domain, which, with its notion of potentiality, provides a vocabulary for dealing with this vexing problem of the mixt or compound versus the constitutive element. The next two chapters deal with the philosophical clash between chemistry and physics, not interpreted as two different academic disciplines but as two different approaches to matter. Thus, the interest of chemists is in the property-bearing principles that animate the chemical reactions they provoke, observe and instrumentalize in their laboratories, while that of the physicists is the ultimate causes that lie hidden behind the sensible phenomena. This confrontation, expressed in terms of elements versus atoms leads us to a consideration of Mendeleev's conception of the element, which is a key historical and philosophical concept in chemistry.

Treating Mendeleev allows us to address the issue of reductionism, which becomes particularly important following the introduction of quantum mechanics early in the twentieth century. Indeed, Mendeleev's epoch-making periodic table reflects a subtle abstract philosophical understanding of the element that was severely put to the test by the discovery of isotopes and the birth of atomic physics. In Chapters ten and eleven, we address the question of positivism, as chemists have often been dismissed as naïve positivists, particularly when they stubbornly refused to accept the existence of atoms in the nineteenth century. We will explore the tradition of positivism from Comte to Mach and show how chemistry both is and is not positivist in its approach. Two chemist-philosophers, Ostwald and Duhem allow us, nevertheless, to see the limits of certain forms of positivism in the context of chemistry. The discussion of atomism raised by positivism also allows us to explore the variety of atomisms that have existed and continue to exist. Thus, the periodic table embodies a distinctive chemist's atomism focused on the atom as a node of chemical relationships.

Chapter twelve represents the clearest, most direct presentation of our own philosophical position, which we term "operational realism". We regard this as being implicitly or explicitly the characteristic philosophical stance of the chemist in the laboratory. The attention that chemists pay to the specificities and idiosyncrasies of the chemical materials they have in their hands or manipulate with the aid of instruments is what sets them apart from theoretical physicists. This should not, however, be used as an argument against the philosophical legitimacy of chemistry, which neither addresses the "essential" questions nor seeks *the* unified theory, but rather as a basis for rethinking the terms of philosophy of science. Thus, we offer a strong argument in favour of a reconsideration of the problematic of the philosophy of science with chemistry occupying the place it merits.

We close the book with some reflections on nanotechnology and the profound transformations that the past few decades have wrought on chemistry and its neighbouring disciplines. Indeed, we pose the question of whether nanotechnology marks the end of chemistry as a discipline. Whatever the ultimate response to this question, it is impossible to deny a strong continuity in the relationship between science and society in the transition from chemistry to nanotechnology. A re-examination and renewed interest for the natural at the level of the nanometer has also seen the resurrection of the Faustian ambitions associated with chemistry in the past. Nanotechnology seeks not only to mimic nature, but also to outdo it, with increasing numbers of scientific visionaries heralding the mastery of artificial life and self-propagating nanomachines.

In Chapter fourteen, in guise of a conclusion, we turn to consider the ethical issues raised by the whole history of modern chemistry and how these might be addressed as we move into the era of nanotechnology. We suggest some general philosophical guidelines that could help to structure a new ethics appropriate for contemporary research in the context of the ongoing nano-revolution.

References

1. J. Simon (2005) and A. Donovan (1993).
2. D. Baird *et al.* eds (2006), J. Schummer (2003), J. van Brakel (2000).
3. A. Donovan (1993).
4. In particular, we recommend B. Bensaude-Vincent and I. Stengers (1996).

CHAPTER 2

CHEMISTRY AND POLLUTION

Today, most people automatically refer to something as being either "natural" or "chemical", with this dichotomy conceived as being both exhaustive (everything is either chemical or natural) and exclusive (something cannot be both chemical and natural). But you do not need to find a particularly pedantic scientist to remind you that all natural objects and materials are composed of chemicals, as is all matter, and that many products, such as aviation fuel, for example, that we talk of as being "chemical" are simply purified "natural" substances. Nevertheless, to dismiss this dichotomy as groundless, an approach sometimes adopted by chemical companies in their communication strategies, is to miss the deep sociological or psychological message that lies behind this widely accepted distinction.

Chemical versus Natural

The dichotomy between natural and chemical translates the lasting impression left by a little over a century and a half of the heavy chemical industry and its increasingly prolific synthetic offshoots. In its short history, the chemical industry has generated more benefits and fears than any other sector, except perhaps nuclear energy. Most of us live in a state of profound contradiction with respect to this industry. While it is difficult to imagine life without the convenience of household bleach, air freshener, insecticides, specialized rubber soles for our shoes and resistant nylon materials, it is with unease, if not alarm, that we observe waste sites filling up with undegradable material, and learn about potentially poisonous products and by-products leaching into our soil and polluting our fresh-water systems, perhaps forever. When a rapidly growing hole was detected in the ozone layer, it was the chlorofluorocarbons (CFCs) produced by the chemical industry as aerosol propellants that were held responsible. The chemical industry is also particularly visible when it comes to counting the cost of globalization. While the imperative to reduce costs by delocalizing dangerous or expensive industries was at the base of the disaster in Bhopal, where thousands of Indians were killed or maimed following a leak from a Union Carbide installation; it was somehow highly appropriate if not entirely predictable that it should be a chemical plant that brought death to a defenceless community in a developing country. Thus, the chemical industry has left a bad taste in the mouths of a public that seems as ready to condemn this industry as it is reluctant to recognize the enormous benefits it has reaped from the same source. The millions of dollars spent every year by chemical companies on advertising and communication campaigns do not, however, seem to be able to erase this negative perception by their presentation of a positive one. Why is this so? In part, it is because the chemical industry has been and continues to be responsible for considerable pollution of the environment, both admitted and unacknowledged. Furthermore, thanks to the prodigious capacity of chemists to produce new products and the dearth of resources available for research to determine their potential hazards, let alone those of combinations of such substances, there are always new potential threats on the horizon.

Part of the negative image of the chemical industry is also due to the poor image of chemistry as a science, a topic we will be addressing from a historical perspective in the rest of the book. This vision of chemistry as a lesser science than physics, for example, makes it natural to categorize it as a technologically oriented applied science, or just the kind of science that would be responsible for the degradation of the environment. Furthermore, as we have already suggested, there is a very real history of environmental pollution by the chemical industry.

The association of the chemical industry with pollution is not something that started only in the 1970s, it is as old as the chemical industry itself. In the nineteenth century, struggles between local populations and polluting dye factories, for example, were successfully kept under control thanks to a robust faith in the potential of scientific and technological progress. Prior to the Second World War, the Faustian deal struck between society and its scientists seemed worthwhile, with the unlimited gains promised by science more than outweighing the limited risks involved. At the height of this scientific optimism, the giant chimneys of chemical plants billowing out black smoke were used to represent not only a nation's prosperity, but also its degree of civilisation. (Figure 1)

This official optimism with respect to the redemptive power of progress never completely succeeded in drowning out the protests of those who did not appreciate the destruction of their environment by new chemical

Figure 1. Illustration by Leon Soderston for the book by A. Cressy Morrison *Man in a Chemical World: The Service of Chemical Industry* New York, Charles Scribner's sons, 1937. Private collection.

pollutants. Imagine the shock experienced by people living in small towns or villages in the Provencal region of France or in Lancashire, Britain when an alkali works opened up. The need for abundant water supplies meant that the production plants were situated near streams or rivers, and often in essentially rural regions. The effects were immediate: not only did nauseating smoke start billowing out of their chimneys into the air, but the population grew with amazing rapidity, generating overpopulated workers' districts filled with sick and suffering humanity. Furthermore, the water sources that attracted the chemical works in the first place quickly became polluted, as little or no effort was made to clean the water that was returned after being used in the manufacturing process.

It is one of the ironies of the history of chemistry that the initial protest against the pollution of the environment by chemists surfaced at the same moment when they were not only constituting a professional body specialized in issues of hygiene and food safety, but were also behind improved agricultural yields and life-saving innovations in the pharmaceutical industry. Indeed, the new chemical fertilizers were hailed by many as the means for putting an end to hunger in the world. In England, local protests against industrial pollution began as early as 1830, becoming a National political issue by the 1860s, and giving rise to the Alkali Act that imposed certain restrictions on chemical manufacturing in 1863.[1] A few decades later, the synthetic dye works that were set up along the banks of the Rhine and the Neckar came into conflict with local farmers and fishermen, due to the toxic effluents that were being dumped into the rivers. Despite the damning official reports on the amount of acid and other solid wastes found in the water, these controversies remained local and did not lead to any fundamental debate over the desirability of the implantation of chemical factories, or even the responsibility of their owners.[2] The yellow and green rivers seemed less dramatic in light of the profits flowing from the mass production of the synthetic dyestuffs and other modern chemical products. Companies like BASF became large enough to counter any such protests, particularly as an increasing proportion of the local populations owed their livelihood to the factories. Such conflicts of interest continue to mute protests even today, and form an integral part of industrial capitalism's long, troubled history with the environment. Nevertheless, even if protests are defused by concern about local employment or a fear of legal retribution by powerful multinational companies, this only adds to the sense of frustration and mounting resentment that feeds a widespread hostility to chemistry on the part of the general public.

Chemistry in Literature

If we search for chemists or chemistry in Western literature, it is difficult to find any representations of truly modern chemists, let alone positive ones. As both Haynes and Schummer have shown in their analyses of the representations of scientists, the recurrent figure in this literature — particularly in the nineteenth century — is of an old-fashioned, mystical savant, closer to the ancient alchemist than the modern-day chemist.[3] This persistence of the figure of the alchemist reveals a widening gap between the realities of industrial chemistry and the timeless Promethean figure of the chemist himself. From Goethe's Faust to Shelley's Frankenstein, the chemist is presented as a kind of powerful yet deluded (if not insane) magician who tries to rival Nature herself by playing with the dark forces of the natural world without ever truly mastering them. In general, the chemist serves as the allegorical figure for Hubris or man's original sin of pride. As with Faust or Frankenstein, the scientist is prepared to make a pact with the devil in order to be able to play God on Earth.

In his novel, "*Joseph Balsamo, Memoirs of a Doctor*", Alexander Dumas compounded the excesses already associated with the chemist when his protagonist combined hypnosis with chemistry to allow him to manipulate his victims' minds. Overall, the image that persists in the popular imagination is one of mystical alchemy rather than modern chemistry, even though over the course of the twentieth century, chemistry has established itself as a respectable department in most modern universities. In novels, chemistry is almost always represented as a strange, foreign science, with its origins in distant lands, whose importation into our society serves only to destabilize it.[4] While other collective sciences are sometimes seen to contribute to the public good or at least to economic growth, the chemist continues to be depicted as a solitary researcher consumed by his passion. Chemistry becomes an obsession that keeps the adept removed from society, even on the rare occasions when the chemist is on the cutting edge of the science and not lost in its mystical past.

Even when chemistry is presented in a favourable light as a booming modern science, the material benefits it has provided are generally considered to have done more harm than good because they have contributed to the erosion of traditional spiritual and religious values. Just like the alchemist's gold, the wealth amassed by the chemical industry is considered to be immoral and ill-gotten gains that disrupt the economic and social orders founded on honest, productive labour. When some chemical product is identified as being a health risk, there is almost an instinctive reaction of "I told you so". Indeed, it is rare to find authors who discuss, let alone accept the Faustian bargain represented by the modern production of synthetic chemicals. Richard Powers's novel, "*Gain*", provides a notable exception. It is written as two parallel stories, of which the first describes the rise of a large diversified chemical and later pharmaceutical company, Clare, and the codevelopment of its company town Lacewood, Illinois. The second story recounts the illness and death of Laura Bodey, a divorced mother of two and resident of Lacewood who develops ovarian cancer. When a group of patients tries to launch a class-action suit against Clare, Laura Bodey refuses to participate.

> Sue them, she thinks. Every penny they are worth. Break them up for parts.
>   And in the next blink: a weird dream of peace. It makes no difference whether this business gave her cancer. They have given her everything else. Taken her life and molded it in every way imaginable, plus six degrees beyond imagining. Changed her life so greatly that not even cancer can change it more than halfway back.[5]

Nevertheless, it requires remarkable lucidity to make this kind of cost-benefit analysis when faced with human tragedy, and it is not clear that the author would be as understanding if it were he and not his protagonist dying of cancer. In a similar way, the story of Faust has various endings. In some versions, Mephistopheles having kept his part of the bargain carries away the unfortunate doctor to hell, while in others Faust repents in the face of his imminent damnation and his soul is redeemed. We do not yet know the ultimate outcome of our modern Faustian pact with chemistry, but a number of commentators have already predicted the worst.

Silent Spring

Rachel Carson's book, *Silent Spring*, drew attention to a mounting ecological problem in post-war America. Its serialization in the *New Yorker* alerted John F. Kennedy to the dangers of pesticides and herbicides, leading to the constitution of a commission to investigate the problem. The book, which recounted the hecatomb of North American fish and wildlife due to ill-conceived campaigns to tame nature using synthetic chemicals, enjoyed unparalleled public success that brought the issue to the attention of journalists, politicians, and business leaders alike. Indeed, it was only after a series of technical articles denouncing the environmental effects of DDT failed to generate any reaction that Rachel Carson mobilized her popular literary skills to produce a generalist book intended to touch a broad public.[6]

"Silent Spring" opens with a futurist fiction, presenting an American town where the wildlife has been eradicated, where no more fish swim in the streams, and no more birds sing in the trees to herald the coming of a new year; hence the title. The narrative strategy is to juxtapose a bucolic vision of the countryside with a post-apocalyptic image of the world after it has been visited by modern chemical products. The ensuing calamity is of biblical proportions, akin to the plagues brought down upon Egypt by Moses. The desolate and lifeless town described at the beginning of the book represents an extrapolation of the consequences of the continued indiscriminate large-scale use of the new herbicides and pesticides, such as DDT and other organophosphates.

While much of the appeal of Carson's book lies in its mix of evocative imagery and scientific information garnered over decades from journals and colleagues, its force owes a lot to the anthropological dimension of her argument. In Chapter 2, entitled "The Obligation to Endure", she places the relationship between humans and nature firmly in the foreground and presents chemistry as having launched an all-out war on nature. Carson contrasts the quarter of a century of man's destruction of the environment with the long history of life of earth, underlining the fact that *Homo sapiens* is the only species that has ever sought actively to eliminate another. Carson goes on to emphasize a paradox behind the US government's approach to pest control at the local and Federal levels. She argues that the remedies invented to defeat more or less minor problems (from hedgerows that obscure a driver's vision on twisting country roads to Dutch Elm disease) are capable of causing far worse damage than the ills they are meant to combat, and rarely, if ever, achieve their stated goals. This argument resonates with a series of images that have attained mythical status in Western culture. First, there is the ancient Greek concept of the *pharmakon*, both poison and remedy, which was already associated with modern chemistry during the First World War when, depending on the vagaries of the wind on the battlefield, poison gas deployed against the enemy could equally well end up killing the troops who were using it. As with the use of poison gas in the war, Carson suggested that the large-scale use of DDT could easily end up compromising the health and even the civilisation of those who use it. Although intended to limit its effects to killing only "harmful" insects, the principal characteristic of such chemical weapons is their capacity to kill indiscriminately. Carson argued that the problem of modern chemical profligacy was not any particular malevolence on the part of the chemical industrialists (despite what some conspiracy theorists might have thought) but their rank ignorance concerning the potentially harmful consequences of the widespread dissemination of their products. This ignorance, which was linked to a public display of unbounded optimism, duped the population into thinking that those responsible for such practices knew what they were doing and were capable of foreseeing and averting any negative effects. Indeed, at her most charitable, Carson seems to suggest that the government agencies responsible for much of the DDT spraying were equally blind to its obvious ill effects. Nevertheless, the chemical companies were guilty in her eyes of presenting only "half truths" and sugarcoating the unpleasant reality of their products.

The second pervasive mythical image that Carson mobilizes is that of the opposition between cosmos and chaos. The classical vision of cosmos represents an orderly, harmonious world, while that of chaos is an uncontrollable disorder. For Carson, the state of nature represents the cosmos (harmony, balance, equilibrium) while human civilisation with its concentrations of built environments and population is presented as a rupture of this order that triggers uncontrollable chain reactions plunging the world into chaos.

"*Silent Spring*" presents a Manichean vision of the world: nature is seen as a harmonious paradise, while chemistry is cast in the role of a malevolent power, an unbridled force waging an undeclared war on nature. But Carson takes care not fall into the trap of the anti-science movement, calling instead for a new ecology that is at the same time both scientific and political. While she believes in science, she wants to see an alternative version put in place, one that leads humanity down the path of genuine progress by substituting long-term solutions for the contemporary strategy of short-term fixes. What Carson argues is that we cannot hope to control living organisms without first trying to understand them in all their complexity. While the greatest error of the pesticide campaigns of the 1940s and 1950s was the indiscriminate toxicity of the products used, it was coupled with another, more profound error. This error was to imagine that the targeted "pest", whether the beetles that transport Dutch elm disease or malaria-carrying mosquitoes, could be treated quite independently of other species. Thus, Carson proposes that more effort should be made to study the various equilibriums between different populations of animals and their environment. While such studies were being undertaken by a number of university biologists, they were still in their infancy and appeared not to be of any interest to those charged with managing the environment. Carson also wanted to attack the technological fatalism that orientated wildlife and environmental management in the United States. The philosophy behind the mass spraying of DDT and other chemicals was that scientific progress necessarily brought with it improved modern technologies that worked better than other traditional approaches, rendering its proponents blind to the potential negative effects of these modern techniques. Carson suggests that America needs to take another route, that of "good science", which seeks out biological solutions based on a thorough knowledge of living beings and a holist, encompassing vision of life on earth. With this approach, Carson orients her thinking towards the conception of the earth associated with the myth of Gaia, which treats the whole globe as a living being.[7]

We do not intend this critical analysis of the arguments behind Carson's "*Silent Spring*" as a defence of all things chemical or as an attack on the book itself. Clearly, Carson denounced a range of chemicals and associated land-management practices that America is better off without. Our aim was to indicate the range of resources that were mobilized in order to oblige the authorities to introduce a system for the management of these technological risks. In order to alert public opinion, Carson needed to revitalize certain mythological tropes in this new context, thereby creating a modern mythology that opposed chemistry, a new avatar of the forces of evil, to ecology, representing the forces of good. Furthermore, the extent of the public, government and industry reaction to *"Silent Spring"* amply illustrates how effectively this skilful mix of fable and myth served as a motor for political action. The *National Agricultural Chemical Association* perceived the book as such a threat that they launched a quarter-of-a-million-dollar television and print-media campaign to discredit it. On television, a doctor enumerated the victims who would have paid with their lives for the absence of the "poisons" denounced by Carson, while a chemical firm published a spoof of the book parodying her terms of a desolate wilderness. In it, they described a world without pesticides, a world ravaged by starvation and a thousand other plagues. Such over-the-top responses attempting to instil a terror of a world without chemicals could only serve to accentuate the ambiguity of the image of synthetic chemicals as both poison and remedy among the public at large. Their effect was equally contradictory, providing a huge amount of publicity for the original book, and fuelling a popular movement that put even more pressure on the government to intervene. As we have already mentioned, President Kennedy was alerted to the issue by Carson's writing, and the inquiry he launched resulted in the creation of the *Environmental Protection Agency* in 1970, and the banning of DDT in 1972. In the same year, the United Nations organized the first *Conference on the Human Environment*, which was held in Stockholm. Here, activists from around the world drew attention to an environmental emergency, arguing that international environmental policies had to be introduced to preserve the earth's dwindling resources.

Humans, it is often said, learn from their mistakes and nowhere is this truer than in the case of the applications of technology.[8] The errors of conception and judgement denounced by Rachel Carson in her *"Silent Spring"* have provided many lessons that have since been integrated by environmental managers and chemists alike. The most important lesson taken on board by the chemists was to undo the association between chemistry and an unreflective massive intervention in nature; that a chemical is found to kill a troublesome insect is now no longer considered a good enough reason to blanket millions of acres of farmland with it. The resulting reformed method of chemical intervention is represented by integrated pesticide management; an approach in which targets are clearly identified and the use of pesticides is limited with short-action products favoured in an effort to limit any undesirable side-effects. Nevertheless, stepping back and taking a global perspective, we cannot escape the fact that the worldwide use of pesticides has grown continually since the 1970s.

"Better Things for Better Living … Through Chemistry"

The Manichean vision that pitted chemistry against nature in Rachel Carson's "*Silent Spring*", an image that still seems to be favoured by certain ecologists, arose in a particular historical context that is no longer our own. Carson was reacting to the massive intrusion of synthetic products into the Western world. Since the 1950s, this initial incursion has become an invasion which has proved so successful that synthetic chemicals are now ubiquitous. Indeed, their omnipresence has had the paradoxical effect of making synthetic chemicals almost invisible and it takes a certain amount of historical reflection to bring them back into focus. Take a simple example; when was the last time you consciously noticed a Formica kitchen surface?

The prosperity of the chemical industry after the war depended to a large extent on the mass manufacture of its products. Such mass production in turn required the creation of mass markets for these products, leading to the use of advertising techniques that often challenged traditional social or cultural values. On 15 May 1940, for example, a giant, two-ton model of a woman's leg encased in a fine nylon stocking was erected in Los Angeles. In the days that followed, thousands of shoppers queued up to get their first pairs of nylon stockings, which a clever advertising campaign had transformed into the symbol par excellence of modern feminine luxury.[9]

We should not forget that nylon was launched into the American market while Europe was plunged into a war that would cost the lives of millions of men and women. Furthermore, while chemicals had already been used in combat in the First World War, notably the phosgene and mustard gases deployed on the Western Front,[10] science was more deeply and directly implicated than ever in the Second World War. After the war, the story came out that the German chemical companies had collaborated to produce Zyklon B. This deadly gas was used by the Nazis in the gas chambers of concentration camps as part of their modern genocidal campaign against the Jews and other targets of their policy of extermination. Although doubtless overshadowed as an illustration of the destructive potential of science and technology by the explosion of the atom bombs over Hiroshima and Nagasaki, this story of Zyklon B particularly marked the chemical industry. Even before the Second World War, however, chemical companies were working hard to distance chemicals from images of death and destruction evoked by poison gas and explosives. Probably, the most famous slogan in the history of the chemical industry "Better things for better living ... through chemistry" was launched by DuPont in the 1930s with the aim of countering the negative association that had formed in the public mind linking chemistry to warfare.[11] DuPont's campaign went much further, however, creating an image of a chemical industry that could provide a new, better way of life through its creation of boundless material abundance, thereby heralding a new age of universal prosperity. Even the staunchest opponents of the modern era of consumerism have to admire the genius of the publicity agents charged with the daunting task of marketing 6-6 polyamide, a new synthetic fibre developed in the 1930s by a team of chemists at DuPont under the direction of Wallace Carothers.

Although nylon was far from being the first synthetic polymer, the way it was marketed constitutes part of an important development that binds the history of chemistry to the history of our modern consumer society. Thanks to the history of these polymers, which continues right into the twenty-first century, we can trace a series of complex yet highly significant transformations in the values attached to both the artificial and the natural. Celluloid, the first synthetic polymer to be manufactured in bulk, was produced by John Wesley Hyatt in 1870 and was used to make billiard balls. There are several lessons to draw from this early episode in the history of synthetic polymers. First, the initial use of celluloid was to make an object traditionally manufactured from a naturally occurring but rare material, in this case ivory obtained from elephant tusks. Thus, this is not a case of the imitation of a "natural" object by an artificial one, but rather the substitution of the material used to produce the same artificial object. Second, celluloid billiard balls were considered inferior to ivory ones, because, while they were cheaper and looked similar, they did not possess the properties that made ivory billiard balls good for playing billiards.[12] This trade-off, which consisted in saving money in return for a product of lower quality, has been a constant element in the perception of "plastics" throughout their history. Indeed, the idea of plastics being cheap substitutes has remained in the collective consciousness, even now when polymers are often the material of choice for a given purpose, and can easily be much more expensive than "natural" materials. Nevertheless, no one would think of talking about wood as a "substitute" for a high-priced synthetic polymer, even less of considering iron as a low-quality "imitation" of a plastic.

In 1907, the Belgian chemist and inventor Leo Baekland combined phenol and formaldehyde to create "Bakelite" a plastic of entirely artificial origin. The material offered a number of commercial advantages over celluloid as it was easy to mould and kept its shape even in quite extreme conditions. It was also an excellent electrical insulator. When it was launched in the 1920s, Bakelite was not promoted as a low-cost imitation of naturally occurring substances, but was instead actively associated with the theme of abundance. It promised a mass-culture of opulence, making luxury items available to one and all.

The promotion of polymers also required overcoming the negative image of their non-specificity or their adaptability to a wide variety of uses. Indeed, what is considered today to be a key advantage of such plastics was originally seen as a serious drawback. While each different natural substance seemed to be suited to only one or a few purposes, Bakelite could serve many different functions, and was used to make combs, hair slides, collars, buttons, telephones and radios to cite only a handful of its better known applications. Furthermore, the possibility of dyeing bakelite meant that it could be given the appearance of a variety of different materials, such as tortoiseshell, amber, coral, marble, jade, or onyx. Like its polyvalence, its chameleon-like ability to take on different colours devalued Bakelite in the eyes of the public, as they were seen as clear indications of the radical inferiority of the artificial substitutes with respect to the original natural materials. Over half a century, however, a radical change of values occurred in the modern Western world, and these defaults of plastics were transformed into its leading qualities. The historian of American culture, Jeffrey Meikle, has published a perspicacious analysis of this shift in the US, explaining it largely in terms of the technological advances that allowed manufacturers not only to enhance the appearance of the materials (so that they looked less "plastic"), but also to enhance their properties (mechanical strength, fire resistance, stability over time, etc.) with respect to their diverse uses.[13] Meikle also cites the influence of a series of advertising campaigns presenting arguments that while familiar to us today were quite novel at the time. These advertisements advanced two types of argument to valorize these synthetic polymers in the eyes of the public. First, there was the environmental argument stating that the use of synthetic chemistry saves valuable natural resources (thus, synthetic fur, for example, saves the lives of foxes or baby seals), and second, the industry tried to convince the consumer that their synthetic products were actually better than the natural material they replaced. These chemicals answered to standardized norms and could be guaranteed one hundred percent pure. A product without any defects, the argument goes, should be considered superior to one made out of variable natural material, which, because of this variability, was inherently unpredictable if not unreliable.

Leaving aside individual advertising campaigns, the widespread adoption of synthetic materials, particularly in the production of textiles, was the principal factor behind a radical re-evaluation of these substances and the very notion of plastics. In the span of a single generation, the deficiencies of plastics — their protean adaptability in terms of properties and form, their low density, their low manufacturing costs, and even their apparent physical imperfections, like fragility, brittleness or malleability — were transformed into their greatest advantages over their "natural" rivals. The very name "nylon" illustrates that this transformation was well advanced by 1940. Rather than Silkon, Silkex or Silkene, which would have evoked the use of the material to replace silk, DuPont's marketing team chose a word without any apparent link to this natural substance. This entirely synthetic material was to be promoted on its own terms rather than as a cut-price ersatz.

This change in attitude concerning these materials reflects a transformation in the domestic environment, with more and more household items made of synthetic polymers. Furthermore, these materials assumed an increasing range of functions, and could be adapted to all sorts of needs, generating a wealth of products ideally suited to the new American domestic social type: the suburban housewife. The term plastics and the idea of plasticity are synonyms of adaptability, materials that change their form in reaction to someone's needs or desires. This characteristic flexibility has been exploited by designers, architects, and artists alike who have used modern materials to explore new aesthetic horizons, from moulded bumpers on cars to acrylic paintings on canvas. The adoption of synthetic polymers in the highest ranks of the worlds of art and design lent them a certain degree of respectability and licensed new orientations in the design of consumer goods. It was no longer necessary to dissimulate plastics using pastel tones or imitation tortoiseshell finishes. They could now be dyed in bright primary colours, proudly announcing their nature as artificial products of the modern chemical industry. Plastic objects started to be presented to and displayed by the consumer with an insolent attitude, proud to be artificial and no longer struggling to hide the fact that they were not natural. We can see this quite clearly in the advertisement for Plexiglas from 1939 which vaunts the sanitary appearance of the glass-like material without the drawbacks of glass kitchenware (Figure 2).

On the other side of the artificial-natural divide, the specific use of certain natural materials started to be interpreted as a limitation or a mark of

Figure 2. Advertisement for Plexiglas®, a brand-name synthetic polymer made by Röhm and Haas from 1939. With all the advantages but none of the disadvantages of glass, housewives could only look at the resulting affordable hygienic kitchenware with admiration. Courtesy of EVONIK Industries AG, Corporate Archives, Darmstadt.

their rigidity. The fact that these materials were well adapted to only one particular function began to be interpreted in negative terms as a lack of flexibility. It is no longer something positive but a lack of flexibility. Plastic has come to be seen as throwing off the constraints and oppressive weight associated with traditional materials. Furthermore, following the advent of the information age, it is synthetic materials that are seen to most nearly attain the ideal of virtuality. Associated with novelty, change and innovation, the use of plastics became a symbol of precisely these qualities. In his analysis of contemporary myths written in 1971, the French philosopher Roland Barthes described the new relationship to nature precipitated by the dissemination of plastics, qualifying it as miraculous.

> More than just a substance, plastic is the very idea of infinite transformation. It is, as the vernacular name suggests, ubiquity rendered visible. Indeed, this is what makes it miraculous; the miracle is always a sudden change in nature. Plastic is impregnated with this amazement: it is less an object and more a trace of a movement.[14]

This "miraculous" potential for transformation endows plastics with a virtual quality that resonates with an ephemeral way of being that is characteristic of our modern era. Just like the objects that increasingly populate their environment, modern human beings have come to be valued for their flexibility, availability, adaptability, and even their impermanence. The connotations of the word plastic are particularly rich in American culture, with the plastic man as a social type referring us to a happy, superficial, and infinitely adaptable person who is both the product and the producer of the new spirit of capitalism.[15] Even for the anti-globalization movement that rejects the neo-liberal model for the world, flexibility, or plasticity, remains a positive trait.[16]

The resonances of this movement are not confined to politics, however, but have passed into other sciences as well. In the time of large mainframe computers, neurobiologists used to talk of the brain as being hard-wired. Our skulls were supposed to contain a large moist computer of fixed constitution. Today, in the "flexi" age where we can take our laptop computers with us anywhere, everything has become flexi: work-time, holidays, the family, sexuality, etc. It is surely more than just coincidence that the buzzword in neuroscience has also become "plasticity". Today, the brain is seen as a highly adaptable neuronal system that can transform its function according to demand. Nevertheless, our central point is that this plastic age — for better or for worse — was ushered in by chemistry, which is, therefore, indissolubly associated with plastics in the modern industrial world.

As we have been arguing above, plastics played a leading role in the cultural revalorization of the notion of the "artificial". A term that started off implying a low-quality ersatz for natural materials shifted away from this sense of pale imitation and came to represent material desirable in itself. Since then, the artificial has come to supplant the natural in every walk of life. As Meikle so accurately notes, the plastic culture expresses "a faith in technology's capacity for transmuting nature's imperfections so as to arrive at the dazzling perfection of the artificial".[17] Thus, the twentieth century has seen a paradoxical interplay between technology, taste, and culture. While the chemical industry poured out an ocean of artificial products, what might have seemed a recipe for generating negative associations was transformed into a utopian dream of a brave new world. Limitless synthetic polymers are no longer regarded as the raw material for goods destined exclusively for those who cannot afford the real thing. They have come to be seen as the material that perfectly matches the demands of modern life, the only material capable of protecting modern humans against an increasingly hostile environment. Some, if not all, plastic products have come to be objects of desire equalling or surpassing those made of natural materials. Nevertheless, this revolution has not come without a heavy environmental cost.

Abundance and Waste

Synthetic products like nylon and polyester were marketed using the themes of abundance, comfort, and prosperity. Chemistry charged itself with the transformation of society, putting material wealth within the reach of the most humble household. This image of chemistry as the universal provider of luxury goods was particularly appealing during the great depression that followed the wall street crash in 1929. Nevertheless, the problem was that these materials (cheap, light, proteiform) were highly durable and yet part of this economy of abundance. Thus, they were employed in products that were used for a relatively short time, often with obsolescence built in, and yet the material endured long after the objects were thrown away, creating an enormous waste-disposal problem. It is not an accident, therefore, that Jean Baudrillard chose plastics to epitomise the inherent paradox of consumer culture. In this manner, he could point his finger at a society oriented towards the mass manufacture of more and more ephemeral products: "In a world of plenty, fragility replaces rarity as the dimension of absence."[18]

As Meikle has shown, the 1960s and 1970s saw plastics become not only the symbol of modernity and the American dream, but also the target of a critical counter-culture, which associated the material with the materialist, superficial consumer culture it so vigorously condemned. In the same logic, chemistry served as a metonym for a technological, capitalist society that alienated mankind by its mirage of abundance and luxury. Thus, for critics, while being "plastic" may have suggested that a person was flexible and adaptable, it had clear negative connotations: the plastic man was ever-changing, superficial and devoid of any personality. Indeed, plastics epitomize modern, wasteful consumer society to such an extent that they are universally despised and rejected by ecologically oriented movements. Plastic gadgets are the first items to be banished. All clothes should be made from cotton, wool, or hemp, furniture from wood, and tableware from china. The rallying cry of "back to nature" implies the rejection of all synthetic materials as well as the favouring of traditional plant remedies over pure chemical pharmaceuticals.

The spirit of Carson's denunciation has found a wide echo in modern society, with other activists condemning consumer culture in general. The criticisms of the excessive waste associated with post-industrial society have moved the focus from the preservation of national fauna and flora to the management of the whole of the planet. Along with the automobile industry, chemistry is seen as both polluter and profiteer, transforming naturally occurring petroleum products into disposable artificial materials without any respect for the environment or the earth's natural resources. Early on, the ecology movement criticized not only the ephemeral artificial products emanating from the chemical industry but also its profligate use of raw materials and energy.[19] The chemical industry was a particularly visible polluter, pumping toxic waste into the atmosphere and waterways, and, as we have seen, was stigmatized as nature's enemy. Furthermore, the spread of "the polluter pays" legislation has effectively given legal recognition to forms of pollution that are not only harmful to the environment but also pose a potential threat to human health, a situation that is hardly designed to reassure the public.

Courting Disaster

If one considers the last fifty years, it is hard not to associate the chemical industry with serious large-scale accidents. On 10 July 1976, in Seveso, Italy, a reactor exploded at one of Givaudan-Hoffman-Laroche's production plants releasing a highly toxic chemical, a dioxin known by its initials TCDD. Thirty seven thousand people were exposed to this TCDD and 1800 hectares of this area of Lombardy were contaminated.

3 December 1984, a leak of methyl isocyanate from a chemical plant owned by Union Carbide India Ltd in Bhopal, India killed 3 800 men, women and children. Forty people were permanently handicapped with 2 680 suffering less severe handicaps and associated health problems. The precision of these statistics concerning the accident attest to the thoroughness of the subsequent enquiry as well as the numerous legal proceedings that ensued. The media coverage of the twentieth anniversary of the event on 3 December 2004, while limited — often due to ongoing litigation — conveyed an overall pessimism concerning issues of compensation and reparations.

On 21 September 2001, a stock of ammonium nitrate at Total-Fina's AZF plant in Toulouse exploded killing workers at the plant, and causing considerable damage in the surrounding residential area. This accident caused a great deal of anger and suspicion among the local population with respect to the chemical industry. The inconclusive results of the various inquests and expert reports have done little to allay people's fears. The reaction of Toulouse's population reflects the potential dangers associated with the implantation of large chemical plants close to densely populated urban areas. The danger is all that much greater when these plants produce or stock dangerous substances like ammonium nitrate, which had already been implicated in some notorious explosions, and phosgene, which was used as a poison gas during the first world war.

Each industrial accident has taught valuable lessons to industrialists and public authorities alike. The Seveso dioxin accident was the reason behind two European directives concerning industrial safety measures. Nevertheless, chemists know better than anyone else that there is no such thing as zero risk of accidents. The case of the AZF plant in Toulouse demonstrates that even an approved installation boasting a certificate of conformity with ISO norms can explode causing considerable damage. Despite the omnipresence of its products in our daily lives, therefore, chemistry remains a somewhat wild and unpredictable force. The apparently irreducible fact that chemistry poses a threat means that it needs to be strictly supervised and monitored.

Artificial or Natural?

Here, we have presented the case that the dichotomy "either chemical or natural" with all the connotations it attaches to chemistry — poison, stain, pollution — is not simply the result of the credulity or irrationality of an ungrateful public. Thus, the public can very well recognize how much chemistry has brought to them in material terms, how much it has transformed their lives, and still legitimately be sceptical or at least suspicious concerning this science and its industrial applications. We have been arguing that two realities lie behind this opposition between chemical and natural. First, there is the nature of the chemical industry itself which often handles inherently dangerous substances — either as raw materials or as final products. Then, there is the historical context in which the modern chemical industry developed. It grew up within an economy of mass production to supply a new rapidly growing consumer society with plentiful, if not excessive, material goods. The logic of the massive use of pesticides and herbicides is at the confluence of these two elements, involving highly toxic substances distributed in excess in an attempt to radically modify nature, making it more productive or habitable. The nature of plastic products — cheap, mass-produced, and disposable — also made chemistry the science of the vain, superficial, and inauthentic. If we want to engage a reasonable and constructive discussion about the realities of contemporary chemistry, we need to take this historical and cultural context into account. It is not enough for chemical companies to devise more positive slogans and marketing campaigns, they need to address the public's fears — both real and imagined. They can only do this if they take into account the complex cultural roots of the image of the science.

References

1. E. Homburg, *et al.* eds (1998).
2. A. Andersen (1998).
3. R. D. Haynes (1994) and J. Schummer *et al.* (2007).
4. J. Schummer *et al.* (2007).
5. R. Powers (1998), p. 320.
6. During and after the Second World War, Carson worked for the *Fish and Wildlife Service*, where she published a series of popular books on the wonders of marine life; *Under the Sea Wind* in 1941, *The Sea Around Us* in 1951, and *The Edge of the Sea* in 1955.
7. This vision of the earth as a self-regulating, harmonious being, that takes into account a balance of its inorganic and organic components was popularized in J. Lovelock (1979), which was the source of a different current of environmentalism from the one pioneered by Carson.
8. For a discussion of this subject, see H. Petroski (1992).
9. S. Handley (1999), J. L. Meikle (1995), S. Mossman and P. Morris (eds) (1994).
10. D. S. L. Cardwell (1975) and Haber (1986).
11. D. J. Rhees (1993).
12. R. Friedel (1983).
13. J. L. Meikle (1995) and (1997).
14. R. Barthes (1971), pp. 171–173.
15. L. Boltanski and E. Chiappello (2000).
16. C. Malabou (2000), pp. 6–25.
17. J. Meikle (1993), p. 12.
18. J. Baudrillard (1968).
19. K. E. Boulding (1966), pp. 3–14.

Chapter 3

THE DAMNATION
OF THE ALCHEMIST

In the previous chapter, we used the history of chemistry to explore and explain the recent (or perhaps not so recent) crisis of confidence in chemistry. This problem is accentuated by the fact that the crisis resonates with much older traditions and perceptions of chemistry. Thus, our modern-day fear of chemistry reflects a much longer collective memory stretching back at least to the medieval period. Although largely unconscious, this memory is regularly reactivated and accentuated by the media as well as by popular scientific literature and fiction.

Magicians or Charlatans

Sciant artifices, "to the artificers of knowledge"; these are the opening words of a violent attack on alchemy launched by the Persian philosopher and physician Avicenna (980–1037).[1] Translated into Latin, this text was widely read by European philosophers in the thirteenth century. The artificers are, of course, the alchemists; monstrous artisans whose handiwork subverts the order of nature.[2] While alchemy fell clearly into the category of "the arts" in the medieval period, it never rose to the rank of the liberal arts taught at the university. The highly controversial introduction of metals and other chemical remedies prepared in the laboratory into the pharmaceutical armamentarium crystallized the debate around the legitimacy of artifice in general. While the Paris Faculty of Medicine famously banned the internal use of antimony in 1566, the use of chemical medicines, more or less closely associated with the name of Paracelsus, became widespread in the seventeenth century.[3] In general, the alchemists sided

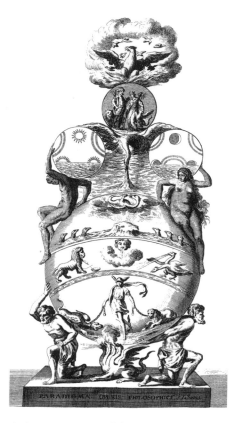

Figure 3. An allegorical representation of the *Opus*, the "great work" of the alchemists. Plate XVIII of the illustrations for the entry on chemistry. Denis Diderot and Jean D'Alembert, *Encyclopédie*, planches chimie, Paris: Panckoucke, 1751–1765. Private collection.

with the supporters of chemical remedies, using it as a pretext to argue for the legitimacy of "technological" artifice in general.

Avicenna's aim in this book was to show that alchemy was a pure exercise in deception. The principal target was the emblematic alchemical practice of transforming base metals (like lead or copper) into valuable or noble ones like silver, or better still, gold. This practice was known as "chrysopoeia," and Avicenna wanted to show that the whole enterprise was a hopeless delusion. To this end, he explained how metals form within the earth and then used two arguments to demonstrate the impossibility of transmuting one into another.

For Avicenna, there is an essential difference between natural gold and its artificial namesake, alchemist's gold. The latter is counterfeit; an imitation of the real thing that merely shares the same appearance. Art being necessarily inferior to nature, no artificial process can possibly change a base metal into another one, let alone a nobler one. This argument of Avicenna's is premised on Aristotle's principle that art imitates nature. In Aristotle's philosophy, the analogy between art and nature turns around the role played by the final cause.

> Art partly completes what nature cannot bring to a finish, and partly imitates her. If, therefore, artificial products are for the sake of an end, so clearly also are natural products. The relation of the later to the earlier terms of the series is the same in both.[4]

Nevertheless, in another well-known passage from his *Physics* II, Aristotle draws attention to a radical difference between nature and art. Natural beings have an internal principle of motion and rest while artificial objects — a bed or a coat, for example — do not possess any such innate tendency to change. Man is born from man, but a bed is not born from a bed.[5] The difference here is an ontological one, which the scholastics reinforced by proclaiming that it was impossible to create an essence or a "substantial form" by art. Thus, while the alchemist's gold might appear to be like natural gold, it will always lack the appropriate substantial form that would make it truly the same.

The second argument proffered by Avicenna concerns the intrinsic properties of each metal. These properties, Avicenna argues, are unknown because they are inaccessible to the senses. The alchemists, he continues, cannot manipulate what they do not know and so are unable to transmute one metal into another. Without the proper knowledge of the properties that flow from the essence of any given species of metal, the alchemists cannot hope to transform it in any way. Thus, apart from any suspicion of charlatanism or downright dishonesty concerning the alchemists' alleged transformations, there were good theoretical reasons for refusing to believe that there could be any other genuine gold but the naturally occurring kind and so dismissing alchemical gold on principle.

The fact that Avicenna developed this philosophical condemnation of alchemists in the context of a translation of Aristotle's *Meteorology* meant that it took on some of the authority associated with the most respected ancient philosopher of the time.[6] Interestingly, *Sciant artifices* became a standard text not only for condemning the art of the alchemists but also for attacking the pretensions of a range of other arts. Thus, farmers who alleged to have created new species of plants found their claims dismissed using the second argument presented above.

Alchemists could find themselves being treated as worse than charlatans, however, with some critics interpreting their practices as a form of magic or even demonology.[7] For Thomas Aquinas, alchemy was neither altogether against nature nor supernatural in the sense of magic, but rather preternatural, from the Latin *praeter naturam*, meaning beside or outside nature. For Aquinas, the category of the preternatural covered a whole range of anomalies — monsters, marvels, prodigies, comets — that while they were not part of the normal course of nature could not really be considered against nature either. Indeed, Aquinas placed alchemy in this category because he was sceptical of the current supernatural explanation that saw demons as being behind the alchemist's art. Whether his art was the product of dupery or magic made no difference to the ultimate conclusion of the Catholic Church, which was to condemn the alchemist.[8] Thus, we can see that the debate raised by the alchemists in the Middle Ages served not only to accentuate the Aristotelian distinction between nature and art, but also to rehearse several of the arguments that are still current today, characterizing chemistry as a science acting against nature. Chemistry was seen to undermine the order of creation including the unchanging order of species, thereby betraying the arrogance of humans who took themselves to be God's equals. In brief, chemistry was against nature. In modern Western culture, the association between chemistry and magic has been reinforced by the Faustian myth about a scientist who concludes a pact with the devil. This myth is based on a real historical figure, an alchemist and astrologist called Johann Faust, who lived at the beginning of the sixteenth century and proudly publicized his exploits in black magic and necromancy.[9]

In Defense of Artifice

The alchemists were, however, quite capable of defending themselves against the assaults we have just been describing. First of all, they could offer experimental demonstrations, using analysis and synthesis to prove (in a quite literal sense) the authenticity of their products. On the theoretical side, they also attempted to demonstrate the legitimacy of their artifice, arguing for an identity between natural and artificial substances at the level of their essence and not just their appearance. Thus, as long ago as the thirteenth century, alchemists tried to impose the idea of man having an authentic power to transform nature, well before Francis Bacon started writing about science and technology, let alone the rise of the mechanical philosophy.

In this vein, a Franciscan monk by the name of Paul de Tarente replied to Avicenna's *Sciant artifices* with a text entitled *Theorica et Practica*, in which he proclaimed that human intelligence gave him dominion over nature and allowed him to manipulate it.[10] Paul de Tarente came to alchemy's defence by using the scholastic distinction between primary and secondary qualities. The arts that work with only secondary qualities — colours, odours, etc. — can never hope to act on the nature or the essence of the substances they manipulate. By contrast, certain arts, like medicine, horticulture and alchemy, which work using the primary qualities — hot, cold, wet, dry — have the power to change the very essence or nature of the substance they act on, allowing them to modify natural species whether for better or for worse. *The Book of Hermes* (the *Liber hermetis*, attributed to Geber), for example, goes as far as to proclaim the superiority of human productions over natural ones. Without challenging the received wisdom that art imitates nature, the author claims that this imitation reproduces the essential nature of the natural substances in the artificial. This is possible because art uses the same means and the same processes as nature, and moreover, has the advantage of being able to improve upon the natural model.[11] Arts such as alchemy are only possible in so far as man can induce nature to produce the things it normally produces in different circumstances. Agriculture is a prime example of an art in which humans make use of natural phenomena — the growing and ripening of grain — in an artificial, more productive context. This argument turns around the

Aristotelian notion of the four causes, so we have to take a little time to explain what they are. For Aristotle, the explanation of any given object relied on four causes: the material, the efficient, the formal, and the final. Using the classic example of the marble statue to illustrate these causes, the material cause is the marble composing the statue. The efficient cause is the action of the chisel on the block of marble that gave shape to the statue. The formal cause is the form of the statue that the artist had in his mind when he carved it, and the final cause is the function the statue is intended to serve: decoration, a fountain, or whatever its function might be. According to Aristotelian defenders of alchemy, the artificial differed from the natural only in terms of the efficient cause, or the agent that produced it. Taking the agricultural example of a field of wheat, we can see that the material cause is the same as for wild, natural wheat, that is to say the wheat itself. The formal cause of the wheat always lies within the wheat grain, and the final cause is to feed animals (in the artificial case it is specifically to feed humans or their livestock). Thus, the only difference is the efficient cause, which in the case of the wheat field is the farmer rather than nature. In 1619, Daniel Sennert published his *De Chymicorum cum Aristotelicis et Galenicis consensu at dissensu* (*On the concordance and disagreements of Aristotelian and Galenic chemistry*) in which he argued that the procedures used by alchemists to transform metals were drawn from nature, and therefore the process was authentic. He also responded to the classic objection that "what nature has united, art cannot put asunder". Those who held this view thought the divorce of the three Paracelsian principles — salt, sulphur, and mercury — even if it had genuinely taken place, was necessarily artificial and could not be natural.[12] In response, Sennert argues, as we have said, that the efficient cause is itself natural, and concludes that "one should refuse the assertion that chemical resolutions are not natural, even if an artisan is involved in his own way. This is because they are achieved by means of fire and heat, and by means of natural causes". Although the chemist's transformations might take place in artificial vessels using man-made ovens, they nonetheless, follow the same pathways as those followed by nature. Thus, the chemist's use of fire and heat is legitimate, because they are just controlled, directed forms of natural processes, generating authentic products.

Through this example, we can see how alchemists and their supporters could remain within the Aristotelian tradition and yet, by attacking the notion of substantial form as well as the ontological distinction between nature and artifice, secure a legitimate philosophical space for alchemical or chemical transmutation. While aiming to demonstrate that chemists could produce artificial bodies identical to natural ones, they also ended up justifying the experimental method and technology itself. This promotion of technology by seventeenth-century alchemists and "chymists" forms an integral part of the contemporary European humanist culture characterized as it was by a faith in man's mastery over nature.[13] Going far beyond an immediate riposte to their critics, some of these thinkers thought they could even imitate God in His ultimate feat, and could use chemistry to create life.

Faustian Ambitions

The possibility of bringing inanimate matter to life has proved to be a tenacious dream for chemists. Paracelsus, for example, believing that sexual relations were too undignified a practice for civilised human beings, attempted to fabricate a "homunculus" by leaving sperm to putrefy in horse dung for forty days.[14] The fantasy of bringing an artificial human to life has been kept alive in Western civilisation through the Jewish myth of the Golem. This fantasy was most famously tied to "modern" science in the nineteenth century by Goethe's *Faust* and by Mary Wollstonecraft Shelley in her best-selling *Frankenstein: or the Modern Prometheus*. In Shelley's story, which was first published in 1818, Victor Frankenstein, a brilliant but overly ambitious medical student, learns how to use chemistry to give life to a body that he has patched together using parts of corpses gathered from the charnel houses and dissection rooms of Ingolstadt. Frankenstein's monstrous creation finds himself to be detestable in the eyes of human beings, including his creator, and takes revenge for his miserable life of rejection and hate by killing Frankenstein's friends and family. This promethean narrative of unbridled ambition and "divine" retribution transposed onto the modern sciences, particularly chemistry (and, more recently, genetic engineering and artificial intelligence), has resonated through the twentieth and into the twenty-first century. It has inspired countless versions of the story, such as *Jurassic Park* and *Matrix*, to name but two of the most successful films on this theme. Indeed, chemists have often nurtured the ambition of creating artificial life, and the resulting projects, which were usually highly controversial, seem to have been animated at least in part by the desire to prove the power of chemistry over the world. Even in the nineteenth century, when such projects were almost exclusively confined to literature, chemists still retained the ambition to compete with, if not outperform nature.

In a book published in 1876, Marcellin Berthelot conceded that a chemist could "never hope to form a leaf, a fruit, a muscle, or an organ in his laboratory"[15] and that his only ambition at this point was to synthesise immediate organic principles, not even as complex as the simplest ones to be found in plants or animals. Nevertheless, like other chemists of his time, Berthelot proclaimed that the synthesis of substances, formerly only obtained from living organisms had once and for all done away with any belief in a vital force that separated living from inanimate matter. Indeed, the idea that Wöhler's synthesis of urea in 1828 had put an end to the notion of the vital force was largely the work of Wilhelm Hofmann and Hermann Kolbe, two chemists who were responsible for some remarkable organic syntheses. Despite the pedigree of those arguing the contrary, Wöhler's synthesis of urea was unable to demonstrate anything concerning the role of the vital force in organic matter because it started from an organic substance, ammonium cyanate, and not the constitutive elements of carbon, hydrogen, oxygen and nitrogen. Although not a total synthesis as defined by Berthelot, the synthesis of urea was an important event in organic chemistry not because it ended all talk of vital forces, but because it raised a new problem for chemistry, that of isomerism. How could two substances apparently composed of the same proportions of the same elements, like ammonium cyanate and urea (CON_2H_4), have different properties?[16]

While only a total synthesis of an organic compound starting from the elements could serve as a compelling argument against vitalism, this myth was, as John Brooke and Peter Ramberg have pointed out, more to do with other issues. Indeed, the very identity of chemistry as well as its future orientation were at stake. What was important for chemists was that the synthesis of organic compounds, hitherto exclusively produced by living organisms, offered the possibility of bringing together organic and mineral chemistry under the aegis of a unified and therefore more powerful discipline. While chemists still had to admit an essential distinction between the "organic" of organic chemistry, and the "organized" of organisms, and so leave the study of the organization and functioning of organs to anatomy and physiology, the fact that substances found in living beings were composed of the same elements and obeyed the same laws as mineral chemistry nevertheless allowed them to establish a foothold in animal and vegetable physiology. It likewise became perfectly legitimate to talk of "applied chemistry" with respect to agriculture, physiology, medicine, etc. Thus, the issues of the power of chemistry as well as the justifiable limits of its territory lie behind many of the scientific controversies that took place in the nineteenth century, such as the debates over fermentation or spontaneous generation.

This desire to expand its legitimate domain is in part what lies behind the identification of chemistry with materialism and atheism during the course of the eighteenth and nineteenth centuries. Chemistry was seen as a disrespectful, if not impious science, motivated by a desire for power, and wilfully neglecting the order of creation, particularly with its pretension to reduce living beings to chemical reactions. While natural history lent itself to the praise of God's creation and provided the principal arguments for natural theology, chemistry seemed like a menace to man's spiritual values and even religion itself.[17] More recent scientists have also argued for the possibility of reducing molecular biology to chemistry. The famous American chemist Linus Pauling confessed that "beginning in 1936, my principal research effort was an attack on the problem of the nature of life, which was, I think, successful, in that the experimental studies carried out by my students and me provided very strong evidence that the astonishing specificity characteristic of living organisms, such as an ability to have progeny resembling themselves, is the result of a special interaction between molecules that have mutually complementary structures".[18] In his 1970 publication, *Chance and Necessity*, for example, Jacques Monod frequently deployed the metaphor of 'chemical machinery' to underwrite his vision of the unity of life from bacteria to human beings.

Whether it be the transgression of the frontier between the natural and the artificial by medieval alchemists, between living and inanimate by their nineteenth-century heirs, or between human and bacteria in the twentieth century, chemistry has challenged the most fundamental classifications that lie behind many of our civilisation's social and cultural values. Thus, even before chemistry came to be seen as a threat to the environment, it was already perceived as a threat to Western civilisation.

From Extracts to Ersatz

Despite the negative attention that chemistry has attracted from medieval times right up to the present, it has to be conceded that the chemical arts have also traditionally represented a form of ancient wisdom, serving to mitigate much of the environmental impact of modern urban life. In the eighteenth century, any large urban centre like London or Paris had groups of women and children scouring the city to collect animal bones or urine for chemical plants usually installed in the suburbs. While these cities were no doubt much more overcrowded and considerably filthier than they are today, they were still also better balanced systems, closer to the "sustainable" model that is so fashionable today. Thus, there was a complex system of recycling that generated the raw materials for a range of "secondary" arts, many of which were chemical. Indeed, it is probably this kind of activity that initially established the famous dictum "nothing is created, nothing is destroyed" so often falsely attributed to Antoine-Laurent Lavoisier.

This work of collecting animal and vegetable products served for a long time as the basis for the chemical arts whose main goal was to render natural products suitable for human use. If we look at the origins of the word "ammonia" for example, we can see the remnants of precisely this kind of process. According to etymological tradition, the word comes from the town of "Ammon" where artisans collected excrement from camels (or other herbivores that fed on salty vegetables), which they then dried in the desert sun to obtain ammonia (ammonium chloride). This ammonia was an essential ingredient of dyes, and served as a finish for metals, as well as being used to make medicines. Likewise, the use of vegetable or animal fibres to make fabric for clothing and other uses was accompanied by the development of a range of chemical arts, in particular dyeing. The colours used in this process were obtained from the three kingdoms of nature: animal (e.g., purple), vegetable (e.g., indigo), and mineral (e.g., azure) by chemical means.[19] While these processes were more or less polluting, the artisans involved only sought to transform the raw material given by nature into a useable form, and did not aim to generate novel by-products whose ecological impact was completely unknown.

Over time, this system of use and re-use of natural resources, which has itself been recycled as an essential feature of sustainable development for the twenty-first century, came to be replaced by another logic. The twentieth century saw economic and industrial rationality raise chemistry up to its highest point, demanding the substitution of natural materials by those produced in the laboratory. This process of substitution has at one and the same time been chemistry's greatest blessing and its most profound curse. In a sense, one can argue that pharmacy was a precursor of this industrial development, as it enjoyed its own kind of "chemical revolution" as early as the sixteenth century. It was in this period that Paracelsian pharmacists tried to add chemical remedies, often composed of metallic salts, to the range of plant and animal extracts that constituted the traditional pharmacopoeia. It was only in the eighteenth century, however, that chemists started to make a wide range of "artificial" salts in their laboratories. The expansion of the chemical arts was further facilitated by the discovery of new procedures for artificially producing essential raw materials like soda ash — known today as sodium carbonate and essential for glass and soap making — or vitriolic acid (sulphuric acid). Indeed, such processes were often at the base of thriving industries that could take off only once the production of these raw materials was liberated from the vagaries of supply from natural sources. Thus, right from the dawning of European industrialization, chemistry was present in the replacement of natural products by mass-produced substitutes, albeit initially of exactly the same nature.[20]

To better understand the issues raised here, it is worth exploring the question of the status of these "artificial" products, using soda ash as our example. Unlike the plastics, one cannot readily talk of "unnatural" materials in these cases. Thus, artificial soda ash (sodium carbonate) differed little from the natural soda ash it replaced, and both were made using naturally occurring raw materials. Natural soda ash was made by burning a plant of the genus *Salsola*, while artificial soda was made using sea salt (sodium chloride) and vitriolic acid (sulphuric acid). In a similar vein, the term "factitious airs" was applied to the gases isolated by the chemists in the second half of the eighteenth century (carbon dioxide, oxygen, hydrogen, nitrogen, etc.) to distinguish them from authentic naturally occurring atmospheric air. Nevertheless, following Lavoisier's investigation into the composition of the air, it was discovered that atmospheric air was made up of several of these gases, which can even be obtained by analysis of "natural" air. The point of this history is not, however, that analysis is more authentic or natural than synthesis but rather that "artificial" and "synthetic" are not synonyms.[21] What is artificial or "factitious" is the result of human intervention in natural processes. The degree of artificiality depends on the number and type of operations implicated in the transformation of natural materials. Thus, while we can conclude that artificial is neither the opposite of natural nor necessarily against nature, the question of the frontier between natural and artificial still awaits an adequate response.

The first point is that we cannot define natural as that which is taken directly from nature as this would be much too restrictive and would also exclude a range of objects that are commonly held to be natural. Too many "natural" products are submitted to chemical or physical operations in order to adapt them to their intended functions. Thus, for example, a natural fabric like wool or cotton, or a dye like madder only appear to be natural in contrast to others that are much less so, but this does not mean that they are just taken from nature and then used. Indeed, the processing of materials like wool or cotton involves highly technological interventions required to separate the fibres, wash them, and prepare them for use.[22]

What we are suggesting is that all domesticated materials are artificial in so far as they are marked by human intervention before use. This is not just a pedantic observation, however, as it is important to consider what is behind the notion of the "natural" in modern society. We would argue that natural is a subjective notion that translates a person's sentiment of how close something is to the non-human world, be it living or inanimate. Wool seems to come directly from sheep and leather from cows, whereas celluloid (a plant-based polymer) is perceived as having no link to any animals or plants. Evidently this perception reflects a reality — the interventions on a cow's skin to make leather do not transform it as radically as the transformation of vegetable matter into celluloid — but the difference between leather and celluloid in terms of natural versus artificial is not absolute, but one of degree. Nevertheless, this difference of degree can have dramatic consequences depending on whether a material falls on the side of the natural or the artificial. While natural has a range of positive associations — comforting, familiar, non-threatening, non-polluting — the artificial tends to attract all the opposite connotations. The fact that the divide between natural and artificial material is not based on any ontological difference is, however, easier to see if we return to consider the early forms of replacement of "natural" raw materials as in the case of the production of soda, as the history becomes considerably more complex following the introduction of synthetic polymers.

Pure and Applied Chemistry

The history of human civilisation has, in a sense, been the history of the artificial. When humans first turned from a life of hunting to one of agriculture, they had already mastered many techniques for preparing skins for clothing and plant and mineral material for building. At some point, wool started to replace hides as the favoured fabric for clothes, representing a prime example of the artificial replacing the natural, although this period of transition has long been lost to the collective human memory. Indeed, the idea that someone dressed in an animal skin could denounce someone else wearing a woollen shawl on the grounds that it was not natural seems ridiculous. Nevertheless, the difference between natural and artificial soda appears to be even slighter.

Is the replacement of the "natural" by the "artificial" simply a question of technological innovation, or is science also implicated in this process? While it is hard to justify the talk of science when considering the introduction of useable wool fabrics, for example, in the case of soda and other products that fuelled the industrial revolution, science was never far away. To illustrate this point, we can examine the role played by chemistry in the introduction of artificial soda. The first role of science was to show that artificial soda could replace its natural competitor in a range of industrial processes. The rise of chemistry as implicated in systematic programmes of investigation and analysis of plant and mineral resources was not only a powerful force behind its acceptance as a respectable academic science, but also led academic chemists to deliver their expert opinions concerning various industrial processes and products.[23]

Henri Louis Duhamel du Monceau, a member of the Paris Academy of Sciences, represents a good example of this new breed of Enlightenment chemist who brought science to the aid of industrial practice. In 1737, he demonstrated the identity of the alkali extracted from seaweed and its "artificial" replacement made using vegetable acids. It would, of course, be wrong to claim that the Leblanc process, which was introduced half a century later, was simply the "application" of du Monceau's scientific result. After all, it took many years of experimenting and tinkering to get the process to work properly and even longer to make it competitive. There is also a famous myth around the man behind the process, Nicolas Leblanc, which presents him as a lone genius pushed to suicide by the state's failure to recognize his invention for its true worth. This legend gives a quite false impression of the French state's interest in such technological innovations, however, as research into the artificial production of important raw materials was strongly supported by the government. Thus, state-sponsored bodies like the Academy of Sciences often organized prize competitions around just this kind of project. Despite how it is sometimes portrayed in popular histories of invention, the artificial substitute material is never an accident. This process of substitution constitutes an important element of an economic logic bound to a culture of innovation that has promoted progress to the status of an absolute necessity. It is very rare that one substance intended to replace another is superior to the original, particularly early in its career, and especially when it is introduced due to a sudden shortage of the first. To succeed in the market, the artificial product needs to prove itself and offer tangible advantages, be they economic, fiscal, physical, or other. Thus, artificial soda gradually became competitive as the new production process simultaneously improved and the economies associated with large-scale manufacture drove the price far below its competitors. This gradual process allowed time for the development of a whole technical system around the production and distribution of soda and other chemical raw materials that fuelled the heavy chemical industry of the nineteenth century.[24]

Artefacts as Hybrids of Nature and Society

During the eighteenth century, governments all across Europe supported the application of the new sciences to areas of vital economic significance, such as mineralogy and medicine, hoping to gain an advantage in the competition between nations that had become so important in this era. In the German states, strongly influenced by cameralism, Joachim Beccher championed chemistry as an indispensable aid to mining and metallurgy and argued that it thereby made an important contribution to the wealth of any state. In Sweden, the state took a direct interest in mining, and its government-sponsored Mining Office served to boost academic chemistry in the country. Johan G. Wallerius, for example, obtained a chair in chemistry at the University of Uppsala by presenting his discipline as the *chimia pura* ("pure" chemistry) that inspired *chimia applicata* (applied chemistry).[25] Chemistry's passage from art to science was achieved on the basis of the tight links between knowledge and action that it promised, although academic chemists often distanced themselves from the direct applications of their theories to practical chemical tasks.[26]

This interweaving of pure and applied science ensured the ambiguous status of chemistry that accompanied its climb to academic respectability in the eighteenth century. Looking at various categorizations of knowledge, we can see that chemistry had largely succeeded in integrating itself into natural philosophy. Nevertheless, examining the classification of knowledge that is placed at the beginning of the *Encyclopédie,* we see that Diderot locates chemistry next to magic and as far away as possible from the other noble sciences that offer the possibility of giving an overview of the world. While physics and mathematics promised to provide an "objective" view that placed the scientist beyond the physical world under consideration, chemistry was resolutely stuck in the world of empirical phenomena. The ambivalence of philosophers did not, however, stop chemistry from becoming a very popular science among a wide public in the eighteenth century. In Paris and Edinburgh, as well as in other European cities, lectures and other public demonstrations of chemistry enjoyed considerable success, whether they were free or demanded an entrance fee. Traditionally, such courses had been the domain of medical students and apprentice apothecaries learning about chemical preparations as part of their training, but increasingly these courses started to attract a bourgeois public with no direct professional interest in chemistry. Perhaps, the best known luminaries to attend such courses in Paris were the philosophers Denis Diderot and Jean-Jacques Rousseau, with the second going on to develop a thorough practical knowledge of the science. From a modern perspective, the Enlightenment takes on the appearance of a kind of golden age for chemistry, a time when it appeared to be an attractive and fashionable science. This positive appreciation of chemistry in the eighteenth century was probably in large part because this was before the emergence of the first big industrial chemical plants in the nineteenth century.

Industry still occupies a predominant place in our perception of chemistry and our assessment of its products. Thus, for example, the adjective "artificial" is much more readily applied to industrial products than those produced in the laboratory. Indeed, the spread of the term artificial corresponds with the transformation from the artisanal to the predominantly industrial manufacture of goods. In chemistry, this transformation coincided with the replacement of batch processes by continuous production methods, a development that took place during the eighteenth century in Northern Europe.[27] In this period of industrialization, then, the distinction between natural and artificial turned around the manner in which chemicals were produced. Although there was no particular reason to think that sulphuric acid produced by a new continuous process was any more artificial than that produced in distinct stages, the process itself was more complex and less readily comprehensible. As Hannah Arendt has argued concerning these new industrial processes, they change the very concept of manufacturing, as the concentration and automatization of production means that the manufacturing process is no longer a means to an end but becomes an end in itself.[28] The development of the chemical industry illustrates this complex transformation even better than the case of electricity used by Arendt herself. The artificial products that have issued from the chemical industry as it put continuous mass production processes into place have fundamentally changed humanity's relation to nature. First of all, these industrial chemical plants polluted the soil, water and air around them on a scale that had never been seen before. Second, as we described in the previous chapter, the artificial products of this modern mode of manufacturing constituted key elements in the development of a modern consumer culture that has supported a century of massive accumulation of objects and other chemical waste. Furthermore, artificial production of chemicals has allowed states to emancipate themselves from the geographical inequality in the distribution of raw materials. The incalculable economic and geo-political consequences of the introduction of synthetic dyes and the Haber process for producing ammonia are just two of the most striking examples of the chemical as political.

Industrialization transformed the economics of production in chemistry as it has transformed every other area of commerce. The most evident effect was to reduce the manufacturing costs for chemical raw materials, but the new production processes also ensured safer working conditions, with economics pushing the factory owners to reduce risks in order to maintain ambitious production schedules. Despite dramatic reductions in price and equally dramatic increases in the quantities of these new "artificial" chemical products that became available during the eighteenth century, they were not perceived as being counterfeit. There was no attempt to deceive the customers concerning the nature of the goods, they were simply substitutes for what nature or international trade failed to provide. Artificial products were at least in part the result of deliberate political choices made by emergent nation states to liberate themselves from the unpredictability of and restrictions associated with international trade in raw materials.

Whether it issues from laboratory or factory, any artificial product is always a co-production of nature and society. There are a number of different ways we can understand this claim, however. First, as was clearly indicated by Gaston Bachelard in his *Rational Materialism*, artificial products, including natural substances in a chemically purified form, are social productions in the evident sense of implicating the intervention of people living in sophisticated human societies. In an experimental setting, a laboratory requires a minimum of researchers, technicians and other assistants to function, while industrial production demands technicians working in quality control, as well as international committees of experts charged with fixing norms and standards. Indeed, even these standards, which appear to be the very model of independent objectivity, are conventions, fixed through negotiations between humans. Second, artificial products, particularly industrial products, owe their very existence to contingent events in the social history of humanity; wars, blockades, economic pressures, fashion, etc. Thus, to cite one example among many, the blockades that cut off the US and Britain from the rubber plantations during the Second World War precipitated an unprecedented effort to produce synthetic rubber on a large scale.

As we have seen with the first synthetic polymers, celluloid and bake-lite, the consumer often looked down on these materials as replacements of inferior quality, even though, or perhaps because they were relatively cheap. Indeed, the scientific and industrial investment required to bring these polymers onto the market were not at first considered positive sell-ing points, and only underlined their distance from the authentic materials they were replacing. Artificial substances were seen as a temporary pallia-tive measure with which the consumers could make do until either they had more money, or the authentic substance became affordable. The German term "ersatz" which was applied to replacements for everyday products rationed during wartime, like chicory used instead of coffee, often served to describe these synthetic polymers. The use of this term sug-gess the tensions created by the introduction of these products whose very nature threatened the divisions between authentic, natural, society and technology.

Nor can the passage from natural to artificial be an instantaneous trans-formation, as there must always be a transition phase. This period serves not only to improve the material and the processes behind it, but also to allow the users to adapt their manipulation and their conception of the product in question. One of the conditions for participating in the culture of innovation is that the artificial product is in constant danger of being replaced itself. Operating on a Darwinian economic model, as soon as the disadvantages (economic, ecological, physical, chemical, electrical, etc.) of the substance in the context of its use outweigh its advantages with respect to a competing material, this particular use is destined to disappear. Thus, the replacement itself is always under the threat of being replaced, and indeed the logic of modern industry and consumer society is that the replacement must one day be replaced.

To conclude this analysis of the notion of "artificial", let us turn once more to the Enlightenment. During this period, the challenge posed by chemistry to the traditional frontier between the natural and the artifi-cial already troubled Jean-Jacques Rousseau, who was also particularly sen-sitive to another crucial frontier — the frontier between nature and society. Rousseau is well known as a champion of nature, the enemy of all things — attitudes as much as materials — artificial, and defender of a phi-losophy in principle opposed to progress in the sciences as in the arts.

Despite this image, Rousseau, who lived at the time when artificial chemical products were starting to replace their "natural" forebears, seems to have understood that the artificial does not so much serve to distance us from nature, but instead redefines nature as a collection of resources that can be improved, if not perfected by human intervention. In his *Discourse on the Origins of Inequality*, Rousseau proposes the fiction of a "state of nature" in order to explore the concept of nature. He is not seeking the immutable essence of nature, but rather a concept that can be deployed — a history. Human nature, like the nature of minerals or the nature of the vegetable kingdom for the chemist, is defined as a source of perfectibility. Just as society allows humans to acquire "artificial" faculties or sentiments (such as pride) that they did not originally possess, it also provides the conditions for introducing new properties into the material world. It is no accident that Rousseau was led to rethink the relationships between the natural, the artificial and society after having consecrated years to studying chemistry and preparing a long manuscript on the subject, his *Chemical Institutions* (*Institutions chymiques*). In this unpublished work, Rousseau presents nature as itself a kind of laboratory furnished with four instrument-elements.[29] Rousseau constructs a parallel between chemistry and society; just as society and culture are elaborated against the background of the perfectibility of mankind, the chemist remakes nature in his attempt to perfect it. This remaking of nature serves in turn to launch society in the endless race of technological progress. In both cases — society and chemistry — the artificial is derived from nature while not being inscribed in it. The artificial does not arise out of any necessity, but rather derives from contingent circumstances and finds its justification in human civilisations; human life as part of society.

From the earliest beginnings of the science, chemists have worked at the limits, and seem always to have wanted to transgress these very limits. Pushed on by a kind of overweening hubris, chemists have advanced the most wildly ambitious projects, often punished — in myth if not in reality — by the gods. These projects have succeeded in introducing many hybrid substances into our world. While the fruits of the chemist's labours have become familiar and so less frightening, they still evoke ambivalent attitudes. Like other objects that blur the boundaries between important categories, these artificial synthetic substances are both fascinating and repulsive. Most of all, they are destabilizing in their capacity to undermine the rules that support our social and cultural systems. In constructing the modern technosphere inhabited by modern mankind, chemistry has inextricably woven nature into the fabric of society.

References

1. For this part of the chapter, the main reference is the work of W. R. Newman (1989) and (2004) particularly Chapter 2. This theme has also been treated more superficially by R. Hooykaas (1972), pp. 54–74. See also P. Smith (1994).
2. Here we use monstrous in the double sense associated with monsters. A monster in the sense of showing something 'montrare' and in the Aristotelian sense of something unnatural, i.e. against the order of nature. See L. Daston and K. Park (1998).
3. See A. Debus (2006).
4. Aristotle (350 BCE), part 8.
5. *Ibid*, part 1, author's translation.
6. In the Middle Ages, it was standard practice to attribute a text to a renowned 'pseudo' author who might have lived centuries before its publication. Furthermore, the term author had quite a different sense from the one it has today. While the author was the source of the 'authority' of the text, there was no sense of any intellectual property or responsibility of the author for the content.
7. L. Daston and K. Park (1998), Chapter 7.
8. After an initial condemnation by the Domincans at the end of the thirteenth century, a papal Bull issued in 1317 by Pope John XXII at the conclusion of an official disputation denounced alchemy as a fraud, a judgement confirmed by the Inquisition in 1396 in its *Contra Alchymistas*.
9. D. Lecourt (1996), pp. 64–65.
10. W. Newman (1989), pp. 433–434.
11. *Liber hermetis*, attributed to Geber, edited and translated into English by W. Newman, *The Summa Perfectionnis of Pseudo-Geber* (Leiden: E.J. Brill, 1991), pp. 11–12. For more contextual information, see W. Newman (1989).
12. This position rehearses a central motor of Boyle's scepticism concerning any of the alleged elements. Were the elements obtained by heating present in the

substance under analysis or generated by the action of the fire itself? The debate over the legitimacy or otherwise of the chemist's experimental proof is not therefore a purely epistemological question, but necessarily raises meta-physical issues as well. We will discuss this issue further in Chapter 5.

13. This humanism associated with the technosciences constituted a principal target for M. Heidegger who was opposed to man's domestication of the world. In M. Heidegger (1954), he argued that technology deeply altered both the aspirations of humanity and its way of thinking.

14. Newman (2004), Chapter 4.

15. Berthelot (1876), p. 271. *La synthèse chimique* was a shorter version of *La chimie organique fondée sur la synthèse*, published as two volumes in 1860.

16. P. Ramberg (2000) and J. H. Brooke (1968), pp. 108–112.

17. It is not by chance that Berthelot openly sided with the critics of the Catholic Church during the controversy over the bankruptcy of science, which took place at the end of the nineteenth century.

18. Linus Pauling quoted in B. Marinacci ed (1995), p. 106.

19. A. Nieto-Galan (2001), G. Emptoz and P. Aceves-Patrana, eds (2000).

20. The classic work on the mass production of these raw materials is A. Clow and N. Clow (1952).

21. For M. Berthelot, the great champion of organic synthesis, analysis is more artificial than synthesis. According to him, synthesis was only meant to reproduce or imitate the immediate principles found in nature. By contrast, a well controlled analysis could allow the chemist to isolate new organic principles that never existed before in any living organism, supplying the missing elements of a series composed of natural principles.

22. As F. Dagognet has pointed out, the use of wool requires a whole range of technological operations — carding, combing, spinning, as well as dyeing mercerising, etc. — 'although we might not have created the material, we have fashioned and improved it so much that it concerns principally us, our technology.' F. Dagognet (1985), p. 104.

23. F. L. Holmes (1989).

24. The Leblanc process for manufacturing artificial soda, taken to mark the beginning of the chemical industry was not immediately considered to be superior to its competitors. It had many teething problems, and the first plants were not only relatively inefficient but also highly polluting, generating large amounts of hydrochloric acid, and sulphur compounds. Thus, like its inventor,

the process had a difficult time during the Revolution, and only took off after Leblanc's suicide, when the salt tax was repealed in 1807, and French production was protected by import duties. Despite these difficulties, or no doubt because of them, the Leblanc process did not stop improving, and many secondary industries grew up around it, either producing the raw materials needed for the process or making use of its by-products.

25. C. Meinel (1983).
26. For a presentation of the distance established between academic philosophical chemistry and pharmacy in France at the end of the eighteenth and beginning of the nineteenth centuries, see J. Simon (2005).
27. For an overview of the industrialization of the production of chemicals, see A. Clow and N. Clow (1952) as well as R. P. Multhauf (1966).
28. H. Arendt (1958), pp. 200–205.
29. See J.-J. Rousseau (n.d.) and B. Bensaude-Vincent and B. Bruno eds (2003).

CHAPTER 4

THE SPACE OF THE LABORATORY

Why is it that chemists are so reluctant to entertain any universal philosophical perspective — to view nature from an Archimedean point located somewhere outside the world? This philosophical distinction reflects a geographic specificity of chemistry — the laboratory. While situated in the wider physical world, the laboratory is a space set aside for productive sweat and toil, for putting the material world to the test by means of a human being's manual engagement with substance. The laboratory, as its name suggests, is above all a place of labour, and this distinctive site of knowledge, invented by the alchemists, remained the exclusive domain of chemists until it was in turn adopted by the other experimental sciences as they emerged. The fact that the laboratory remained the exclusive property of chemistry for a long period of time obliges us to examine this object in more detail, with more than just a suspicion that it might hold the key to our initial question. Thus, it is important to try to understand what kind of knowledge is produced in this characteristic site, and ask what special powers, if any, the laboratory might possess.

If we turn to Diderot's *Encyclopedia* or other dictionaries from the eighteenth century, we are presented with an image of the laboratory as a room packed full of apparatus; but a laboratory is a great deal more than a collection of material objects (Figure 4). It is an environment for executing gestures that have the capacity of informing the assembled materials in a definite sense and thus transforming them into instruments of knowledge.

Classic paintings of alchemists depict them at work in the shadows of already dark rooms, tying them to an obscurity that is both literal and metaphorical (see the cover illustration). The metaphor of this obscurity refers us to the secrecy characteristic of the hermetic arts and the supposed

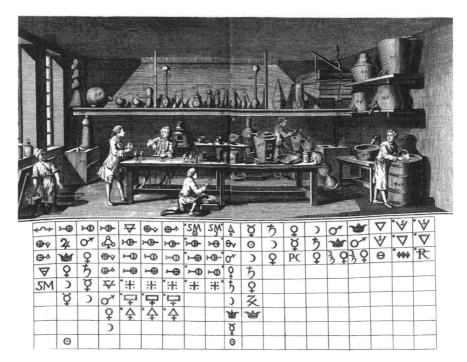

Figure 4. A chemistry laboratory and a table of affinities. The table is an amended and extended version of the one published by Geoffroy in 1718. Denis Diderot and Jean D'Alembert, *Encyclopédie*, planches chimie, Paris: Panckoucke, 1751–1765. Private collection.

magical powers possessed by alchemists, including, notably, the ability to transmute base metals into gold or produce an elixir of eternal life. In the seventeenth century, at the time when many of the stereotypes and myths disseminated to discredit alchemy were being put into place, the laboratory was already in the process of opening up, becoming a lighter and better ventilated place. This did not mean, however, that the laboratory was becoming like the anatomy theatre, the natural history cabinet, or the lecture theatre for experimental physics. The laboratory was not a space for demonstrating knowledge; it was intended neither for putting the laws of nature on show nor for presenting nature in her most intimate details as a public spectacle. While it could be said that the laboratory was a theatre, it was not a theatre for display, but a theatre of operations, a place of transformation. Material is brought into the laboratory to be manipulated and changed into something else; the product that leaves through the door is never the same as the raw materials that enter the laboratory. This transformation also included, traditionally, the alchemist who, by working with the substances in the laboratory, had the power to transform himself along with the material. Likewise, the process that mediated between raw material and end product remained obscure, shrouded in mystery, hidden under the fog of chemical reaction. Against this background, we can begin to understand the symbolic force acquired by the scales, which furnished chemists with a way to monitor what was happening in the hidden space of the chemical reaction. Precise determination of weight gave them a chemical Archimedean point from which to follow and gain power over these transformations. Knowing the weight of the reactants and the products, being able to make a gravimetric comparison between the beginning and end of a reaction — its two fixed points — allowed the chemist to penetrate the black box that formed part of the accumulated empirical knowledge of laboratory transformations.

In order to achieve this kind of control over material transformation, the laboratory has to be a closed, well delimited space protected from the haphazard, complex circulation of material and processes that characterize the natural world. Indeed, this is the very meaning of what it is to be a laboratory, a characteristic paradox that has led to so much productive work in science studies over recent decades. The laboratory is a place deliberately isolated from the rest of the world, and so has little in common with it. Yet, it is a place intended to generate truths about this same natural world.[1] Unlike naturalists who, on the model of the mythical image of Benjamin Franklin standing out in a thunder storm to draw down lightning with his kite, investigate nature in the open, by climbing mountains and walking through fields, chemists work indoors, protected from the elements. Chemists deliberately isolate themselves from natural phenomena to better understand nature. In this respect, the chemistry laboratory serves as the archetype for a strategy adopted by all the experimental sciences. In order to succeed, however, this strategy requires that the scientists make a certain number of commitments with respect to the objects of their investigation. The aim of this chapter is to examine this issue and attempt to identify what might be characteristic of the experimental methods in chemistry.

Chemical Recipe Books

The following protocol is taken from Nicolas Lemery's *A Course of Chemistry*, a very successful chemistry textbook, published in France towards the end of the seventeenth century.

> Slice six or eight ounces of good *Rhubarb*, and steep it twelve hours warm in a sufficient quantity of *Succory* Water, so as the Water may be four fingers above the *Rhubarb*; let it just boil, and pass the Liquor through a cloth; infuse the Residence in so much more Succory Water, as before, then strain the Infusion, and express it strongly: mix your Impregnations, or Tinctures, and let them settle; filtrate them and consume the moisture in a glass Vessel, over a very gentle fire, until there remains a Matter that hath the consistence of thick honey, this is called *Extract of Rhubarb*, keep it in a Pot.[2]

While such "recipes" for preparing the medicines of the time made up a large part of this kind of chemistry textbook, even as late as the nineteenth century, we would not consider this part of chemistry today. Indeed, most modern chemists would consider such a recipe for preparing the extract of rhubarb to be cooking rather than chemistry.

Chemistry has often been compared to cookery, and critics of the science regularly use this parallel as an insult, underlining what they see as a lack of rigorous theory in a domain dominated by experimental techniques. The connection is an obvious one, after all the laboratory is full of equipment (chinaware, glassware, ovens, etc.) that we also find in the kitchen. There is a similar crossover in many of the operations that lie at the base of chemistry, such as heating, maceration, dissolution, grinding up, and crystallization (see the apparatus illustrated in Figure 4). In the mouth of a philosopher or scientist, the parallel between chemistry and cooking is usually derogatory. The implication is that chemistry, like cookery, involves following recipes whose validity has been established by a long history of trial and error, and so implies a work of the hands rather than of the head, far removed from the inclusive explanatory theories that are considered the goal of science. Thus, in many academic circles, chemistry suffers from the contempt reserved for manual occupations. In the context of secondary school, while chemistry is presented as a theoretical science that is principally acquired by means of books, it is generally regarded as less "intellectual" than physics, or rather the image is that the student learns chemistry by more mechanical means such as learning by rote. How many students retain from their chemistry education only the first few elements of the periodic table: hydrogen, helium, lithium, beryllium, boron, carbon? Be that as it may, however much practical work might be involved, the chemistry textbook is the privileged means for teaching the science.

The book has likewise always been the constant companion of chemists in the laboratory. Indeed, the book is often a prominent feature of paintings of alchemists, providing a point of illumination in the ambient darkness. These books are not books of theory, however, they are books containing instructions, collections of recipes for correctly bringing about chemical changes. The first printed chemistry courses or treatises on the subject provided details for preparing medicines, cosmetics, soaps, etc. After some preliminary chapters in which he discussed the elements or principles of chemistry, the author would turn to the essential content; the more or less detailed descriptions of how to arrive at a variety of products. It was doubtless this practical information that usually motivated the purchase of the book in the first place. The book of secrets emblematic of the alchemical tradition had long ago ceded its place to the handbook, *Handbuch*, or *manuel*, which, whether described in terms of recipes (from the Latin *recipere*, to receive) or protocols, contained precise instructions intended to transmit rules and prescriptions for practical action. Thus, we can describe this as a "literature of performance" intended not to construct a discourse, but to teach a savant corporal discipline, to provide gestural rules that were meant to be followed with precision.[3] This language of action, which constitutes the core of chemistry's discourse, is best conveyed when accompanied by experimental demonstrations. Thus, the concept of "paper tools" introduced by the historian of chemistry, Ursula Klein, to describe the manipulation of chemical formulae on paper or in the chemist's head, should first and foremost be applied to the kinds of recipe books or manuals described above.[4] Chemistry cannot, however, be learned just by reading texts. It requires practical experimental work and is learned by doing. Thus, the primary function of the chemistry book is to initiate the student into precisely this kind of practical work, thereby providing a "virtual apprenticeship" or, more commonly, supplementing a real one.

Thus, this literary form of the manual or "recipe book" is an indelible trace of the artisanal origins of chemistry. Constructed out of the practices of metallurgists, glassmakers, soap-makers, apothecaries and dyers. The manual constitutes an essential element of the chemistry laboratory as indispensable as the glassware or china whose deployment it dictates. The importance of these manuals in the correct and safe handling of laboratory apparatus is today disguised by the fact that chemists generally buy their reagents ready-made from suppliers. Although the industrial production of basic chemicals only really took off in the nineteenth century, we can already observe the phenomenon earlier, with Lavoisier, for example, buying many of his raw materials from Parisian pharmacists who were skilled in preparing pure substances.[5] Today, of course, you can order chemicals of guaranteed purity via the internet, and they can be delivered to your laboratory within 24 hours, saving hours if not days of painstaking work isolating, concentrating and purifying the substances yourself. Nevertheless, as every chemistry student knows, during a training in chemistry, you are going to be asked to prepare substances that could easily be bought from suppliers, by following a detailed set of instructions set down in an experimental protocol. When dealing with highly reactive compounds, it is vital to follow the protocol to the letter in order to minimize (if not to eliminate) the risk of explosions or other accidents that nevertheless form a part of the history or mythology of every chemistry laboratory. Unlike a theoretical physicist who works with a pen and paper, or a meteorologist developing a model on a computer, chemists cannot afford to give themselves free rein in their experimental investigations, at the risk of their lives. While deeply entrenched in the concrete, chemistry is tacitly informed by a theoretical framework, an observation that applies even to the pre-modern treatises we have been discussing. Nevertheless, it is so bound up in the particularities of substantial interactions, that chemistry appears not to lend itself to the kinds of (theoretical) approach that would justify its claim to the status of a "true" science.

This specificity of chemistry has given rise to a persistent and deep misunderstanding in the interpretation of chemistry books by historians of science.

It is inconceivable to most historians that experimental practice can stand in for an independent theoretical framework, which is seen as a prerequisite for any physical science. Thus, many are led to suppose that chemistry, unable to generate its own theories, borrows them from its disciplinary neighbour — physics.[6] Hélène Metzger, for example, argued that at the end of the seventeenth century, chemists were "seduced" by the clarity of mechanical philosophy, but after having imported it into their domain found that it did not fit well with the collections of empirical recipes that made up their treatises. More recently, Alistair Duncan has defended the idea that matter theories were rhetorical artefacts deployed by authors in the introductions to their works, only to be hurriedly forgotten in the course of the subsequent chapters, and thus reflecting the fact that chemists were generally practical men concerned only about establishing facts. Uncomfortable with mathematics and reluctant to form speculative theories not solidly founded on experience and observation, chemists were, according to Duncan, satisfied to borrow their theoretical views from "more respectable disciplines". This opinion is similar to that held by Robert Siegfried, who put it this way: "In the eighteenth century, chemistry scarcely possessed anything like what we would today call a theoretical structure, with a coherent central core of axioms, assumptions, or principles from which a significant body of derivative truths could be found consonant with experience."[7] Unfortunately, this common prejudice on the part of historians makes them blind to the conceptual and theoretical constructions that eighteenth-century chemists were able to draw from their practical work of analysis. Thus, as we shall see in the following chapters, vegetable analysis and the experimental study of salts served to transform the notion of the principle/element profoundly. Furthermore, historians and chemists who deny the existence of any indigenous chemical theory tend to interpret any theorization as some sort of attempt to explain macroscopic phenomena in terms of the behaviour of hypothetical microscopic entities. This position is well illustrated by Alistair Duncan in the following citation; "As practitioners of an activity which had hardly completed its progress towards acceptance as an academically respectable branch of philosophy, chemists felt obliged to respect the mechanical explanations."[8]

The dream of chemistry-as-physics was to be able to deduce the properties of any given body from some set of general laws that covered every individual case. In light of this dream, numerous chemists in various periods of modern history have experienced the fact that they have had to follow protocols, manipulate apparatus, handle chemicals, and perform the same actions over and over again as a sort of curse. This sense of frustration is only exacerbated by the experience that these manipulations often end in disappointment, with the experiment going wrong at some point, thereby obliging the chemist to start all over again. One can easily imagine a chemist, exhausted by hours of difficult experimental work, envying a researcher in a deductive science, like mechanics, where predictable calculation had come to replace haphazard manipulation. From this perspective, it is not surprising that there has been a profusion of complaints concerning the chemist's fate. Critics suppose that if one could see each atom, know the position of each molecule, one could (re)construct chemistry as a rational science. But chemists are condemned to stumbling their way through the darkness, trapped at the level of phenomena and never having access to the underlying substantial reality, knowing only the outcomes and not the reasons. As an illustrative example of this posture, we can cite one of the great authorities of early-nineteenth-century French chemistry, Louis Thenard, who, in his *Essay on Chemical Philosophy*, wrote:

> If we could only have precise notions concerning the constitution of molecules and their properties, if we could know with certainty the nature of the force that governs their combinations, then the geometers could submit the diverse phenomena that constitute chemistry to their calculations, and we would then be in the position of being able to elaborate a veritable chemical philosophy. But in the state of ignorance in which we find ourselves at present concerning the intimate properties of molecules and the nature of affinity, how can we aspire to attain the general principles of the science?[9]

According to this kind of analysis, it is the chemist's ignorance that is responsible for the status of theory in chemistry. With chemists incapable of arriving at the first principles for the science or elaborating general laws, they can only aspire to prove tentative, restricted generalizations derived from experimental trial and error. By trying to measure itself against this particular ideal image of a science, chemistry is necessarily going to be found lacking, and will be perceived as imperfect. Indeed, this image of imperfection has dogged chemistry, and continues to do so. How many times have we heard chemists complaining about their inability to solve Schrödinger's equation for the atoms of every element? The difficulty of solving this equation for anything more complex than the hydrogen atom has led theoretical chemists to search for simplifications, and to develop approximation techniques involving fastidious iterative calculations. These more or less *ad hoc* solutions evoke, once again, the crude techniques associated with cookery rather than the precision of the exact sciences, and are usually taken to be just another symptom of chemistry's radical and perpetual imperfection or impurity.

A Space of Toil

Having presented the negative image of chemistry as an empirical and imperfect science, we now want to consider the same characteristics of the science in a positive light. Indeed, the distance between chemistry and the "exact" sciences does not need to be interpreted as a flaw or shortcoming; it can also be seen as offering a privileged position to this science. Some have argued that the very fact of obtaining its knowledge through work and labour guarantees chemistry's autonomy and, more importantly for our argument here, its epistemological originality. Indeed, working, mixing, slaving, and toiling can be considered signs of scientific strength rather than weakness.

This image of the chemist condemned to a life of dirty (and dangerous) manual labour is one of the standard rhetorical tropes associated with the science throughout its history. This image plays out in the context of the wider cultural debate that pits the mind against the hand in the quest to know the world. While many philosophers, particularly in the Platonic tradition, have valued theory and the work of the mind over manual practice associated with the "mere" accumulation of facts, this perspective was far from unanimous. Thus, while the Flemish chemist Johann Baptista Van Helmont suggested that "God sells the arts in return for sweat", the philosopher Francis Bacon contrasted the "ants" who search empirically for the truth with the "spiders" who prefer to spin out theories, proposing the bee as the happy medium appropriate for representing the ideal science.

> Those who have treated the sciences have been either empirics or dogmatical. The former like ants only heap up and use their store, the latter like spiders spin out their own webs. The bee, a mean between both, extracts matter from the flowers of the garden and the field, but works and fashions it by its own efforts. The true labour of philosophy resembles hers, for it neither relies entirely nor principally on the powers of the mind, nor yet lays up in the memory the matter afforded by the experiments of natural history and mechanics in its raw state, but changes and works at it in the understanding. We have good reason, therefore, to derive hope from a closer and purer alliance of these faculties (the experimental and the rational) than has yet been attempted.[10]

In his *Discourse on the Interpretation of Nature* from the middle of the eighteenth century, Diderot resorts to a similar philosophical dichotomy, although he considers the work of the ant and the spider in terms of distinct yet complementary philosophical vocations.

> To gather the facts and to establish the connections between them; these are two very arduous tasks and so philosophers have chosen to divide them up. Thus, some of them spend their lives collecting material, work that is both laborious and useful. The others, proud architects, rush to apply these facts, although until now, time has overthrown practically all of these edifices of rational philosophy. The sooty labourer, as he tunnels on unguided, will sooner or later raise up from under the ground the element that will demolish the architecture formed by the mind alone, and the rubble haphazardly strewn around will be all that remains until another genius attempts to combine it anew.[11]

Diderot clearly favours the sooty empiric over the heady architect, and views the arrogance of the purely theoretical philosophers with disdain, a stance that is echoed in the article on "chymistry" written for the *Encyclopedia* at about the same time by a doctor from Montpellier, Gabriel-François Venel.

Here, Venel defended the chemists' right to cultivate their own epistemological style, a style articulated in a specific idiom. While the chemist's language might well be difficult, dense and obscure, this was precisely because it reflected their unique empirical experience of the world, an experience that was drawn both from the science and the chemical arts. Thus, to understand the special status of chemistry as a sensible as much as a rational science, we need to consider the faculties that are put into play in its empirical engagement with the world. Indeed, chemistry has traditionally mobilized all the senses, generating diverse sensations that have to be coordinated.

Seeing at a Glance

While sight was the sense privileged by philosophers in the eighteenth century, the issue of the relationship between the senses was an important concern. This issue crystallized around several famous enigmas, particularly one which was posed by William Molyneux. In 1688, Molyneux asked whether a blind man who is able to distinguish between a globe and a cube by touch would be able to do so upon regaining his sight, simply by looking at the objects, a question later taken up by John Locke.[12] Furthermore, philosophers were not afraid of relating cognitive skills to faculties of perception, and in his 1749 work, *Letter on the Blind*, Diderot followed Descartes in claiming that blind people reasoned abstractly, like mathematicians.

According to Venel, one of the characteristics of chemists was to rely on their ability to "see at a glance" (*le coup d'œil*). Despite the visual metaphor, this skill implied much more than just vision: "the faculty of judging by a feeling is known to the worker as seeing at a glance, and it is an ability that he owes to the habit he has of dealing with his material." In this way, Venel emphasized the combination of several senses at once in the process of developing an intrinsic and non-verbal form of knowledge characteristic of the skilled artisan. In contrast to the Cartesian faculty of judgement or natural light, this ability of seeing at a glance is not innate. Instead, it is learned through a lifetime of practical experience that breeds practical instincts or intuitions, regarded as being tools of an artisan's trade.

What is special about this kind of knowledge obtained through lived experience is that it is particular and personal. In the case of knowledge derived from reasoning, the reasoning is supposed to force the conclusion no matter who is doing it, making the subject, in principle, interchangeable with anyone else. In this sense, the ability to see at a glance is not interchangeable but specific to an individual. Indeed, in a multitude of specialities, scientific and otherwise, a combination of experience, interest, wisdom, patience or simple obstinacy have allowed certain individuals to attain a *habitus* enabling them to see what others cannot.[13] Thus, to borrow an example from another field, a technician specialized in ultrasound techniques has no problem picking out the heart and legs of a foetus, where the uninitiated just sees a play of light and shadows.

For Venel, having the ability to see at a glance is characteristic of an "artist" in the double sense of the term. First, there is the artist as artisan, the primary sense for the *Encyclopedia;* a worker who, by continual application, has trained his body to the point where he literally embodies a series of techniques that serve him as tools in his trade. Second, there is the artist as creative "genius" a quality that Kant contrasts with mere imitation in his *Critique of Judgement.* The genius, however, is incapable of describing how this creation comes about. The opacity of this kind of "tacit" knowledge echoes Venel's views on chemistry. Thus, when he writes of the foreknowledge or premonition possessed by experienced chemists, Venel once again echoes Diderot's philosophical position.

> The oft-repeated custom of carrying out manipulations gives manual workers, even those in the most unsophisticated domains, a kind of premonition that seems like inspiration. Their failing is that they are subsequently, like Socrates, mistaken concerning the nature of this ability and refer to it as a *familiar demon.* Socrates had such immense experience in judging men and weighing up their circumstances that, in the most delicate situations, he would secretly make an internal calculation that was precise and accurate, followed by a prediction that rarely failed to be correct. He judged men using his sentiment, like people with taste judge productions of the mind. The same thing is true for the instincts of our great technicians in experimental physics. They have observed the operations of nature so often and so closely that they can predict with considerable precision

the course that nature will follow in the cases where they wish to provoke it using the most bizarre types of experiment. Thus, the most valuable service that they can render to those who initiate themselves into experimental philosophy is not so much to instruct them in the processes and results, but instead to instil in them this spirit of divination by which one can sniff out, so to speak, unknown processes, new experiments, and as yet undiscovered results.[14]

This long citation helps us to understand the nature of the *habitus* of the chemist, which primarily concerns sensibility. In general, the experimental sciences have no place for sensibility, but here, sensibility is given a positive spin, being considered as a cognitive faculty, a kind of intuition or talent for grasping reality.[15] In a wider sense, however, emotion, and even passion have their place in a chemist's experimental knowledge, according to Venel. In this context, he cites Beccher's claim that "the taste for chemistry is a madman's passion", and turns this madness into a virtue. The chemist alone has the courage to take on the interminable labour of experimenting, and to see all his time and money swallowed up by his research. Having slid smoothly from epistemological to moral considerations, Venel moves on to the political and the social, describing chemists as "citizens who merit our full gratitude". Thus, we see how in the Enlightenment, experimental knowledge was taken to imply the deployment of emotion and passion, as well as reflecting on the political and moral values of the researcher.

Although contemporary Western culture owes a great deal to the Age of the Enlightenment, we clearly no longer share the mentality of the scientists or *philosophes* living at that time: sensibility, emotion and passion have, it seems, been banished from the laboratory. Experimental science is now conceived of as a formalized, codified process consisting in the formulation of hypotheses, the deduction of material consequences and the experimental testing of the theory. Nevertheless, sensibility has not been entirely ruled out, as we can see in the philosophy of the renowned Cambridge chemist and crystallographer, Michael Polanyi. In his 1958 publication, *Personal Knowledge*, Polanyi emphasizes the importance of "tacit knowledge" for the experimental scientist. This kind of knowledge is non-verbal and is learned only through practice, after which it is incorporated in the head and hands of the individual scientist who thereby acquires a competence that has no formal expression.[16] This conception of tacit knowledge was an element in Polanyi's defence of "free" science against British socialist scientists, like J. D. Bernal, who argued for the centralized planning of useful scientific research. Through the Society for Freedom in Science, Polanyi argued that scientists should be left entirely independent, free to pursue any direction of research. Interestingly, Venel argued for this kind of independence, based on the idea that the judgement of the expert or "connoisseur" depends on their ability to "sniff out" the truth, without being able to offer any explicit justification for their convictions, let alone provide any objective data to support them. As Venel expressed it: "the chemist will orient himself in his normal manipulations, in his daily operations, according to rough and sensible indications, which are always to be preferred because of their convenience, as long as they remain adequate."

A Purified Space

The gestural repertoire of the chemist, no matter how well trained his senses, how fine his intuition, and however passionate he may have been about this knowledge, was insufficient, however, for chemistry to pretend to the status of a science. Nevertheless, Venel demanded the recognition of chemistry as a pursuit worthy of respect in the eyes of the bourgeois readers of the *Encyclopedia*.[17]

How, then, had chemists succeeded in rising to the social status that they enjoyed in the second half of the eighteenth century? One part of the answer lies in the authority they gained from deploying their instruments and other material in order to test and guarantee the quality and purity of a variety of products. Venel himself, like many other contemporary chemists, worked analysing mineral waters. By determining the content of these waters in terms of various mineral salts, the chemists could confirm their therapeutic properties, as well as detecting any potentially harmful impurities. This kind of analytical work, which Lavoisier also carried out when he was a young chemist, requires various measuring instruments, notably the hydrometer (or densimeter), referred to by Lavoisier as "the chemist's scales for fluids". Before the hydrometer can provide reliable measurements, however, it needs to be calibrated based on the comparison of its behaviour in liquids of various known densities, all at the same temperature. The reliability of the hydrometer, therefore, depends upon the skilled use of an accurate pair of scales to provide an independent measure of density, and a thermometer to ensure that measurements are made at the same temperature. In this way, the reliability of any instrument depends on the deployment of others, all brought together to establish an experimental space subject to a strict regime of control. Indeed, it was through his attempts to perfect these measuring instruments that Lavoisier gradually moved from the chemical analyses that formed part of his geological survey work with Jean-Étienne Guettard, to his mature laboratory chemistry. Thus, it was with Guettard that Lavoisier first learned the techniques of gravimetric analysis that would become essential to his new chemistry. Indeed, in Lavoisier's hands, these techniques would transform the laboratory into a theatre of proof, as we shall see in the following chapter. Nevertheless, this kind of analysis is impossible without the essential chemical groundwork aimed at "purifying" the substances to hand.

There are at least two different senses for this concept of purification. First, there is the sense associated with preparing useful minerals from the ores extracted from the mines, or purifying the extracts of medicinal plants to give safe and effective medicines. The apothecaries were traditionally considered the masters of this art, using mild forms of analysis that could separate out a plant's "immediate principles" in their pure state. This was important, as an apothecary's reputation depended on his ability to obtain such pure extracts, and, as a businessman, his livelihood in turn depended on his reputation. Nevertheless, the concept of purity in this context is both provisional and conventional, as it is the chemist's tests and criteria that establish the yardstick of purity, and nothing, in principle, could prevent them from coming under attack.

The second sense of purification involves the determination of the exact composition of a body by separating out its constitutive ingredients. The best example of this form of purification is the practice of elementary analysis that has become a central pillar of modern chemistry. In this case, the quest for purity is pushed even further, involving not only the isolation of the substance in question, but also the removal of all its specificities, the elimination of any contingencies or circumstantial phenomena associated with its origin. The objective of chemistry, as Lavoisier saw it, was to render all chemical substances into generic, unspecific entities characterized by their elementary composition alone.

> The principle object of chemical experiments is to decompose natural bodies, so as to separately examine the different substances which enter into their composition. [...]
>
> Thus, as chemistry advances towards perfection, by dividing and subdividing, it is impossible to say where it is to end.[18]

This process of division and subdivision constitutes a veritable "transmutation" of these natural substances by means of time-consuming experimental work, not without parallels to the alchemical tradition. Substances extracted from nature are subjected to diverse procedures to separate out certain parts, before the constituents are isolated or purified by repeated dissolution, distillation and crystallization. At the end of this process (which, in principle, need never end), the resulting laboratory substance is completely different from the original, inevitably idiosyncratic natural substance. Stripped of its natural history, the circumstances of its production, the techniques that intervened in its extraction, the "raw material" has been transformed into a chemical species; a pure material abstraction. The laboratory substance needs to be "typical" in terms of its characteristic properties, making it unidentifiable in any traditional sense. For example, water used as a solvent in the chemistry laboratory should not taste like Malvern water, Mediterranean water or water from Evian, it should have no taste at all, it should be "pure" water — generally regarded as unfit for human consumption. This transformation changes natural objects into "fetishes" in the sense used by Bruno Latour, artefacts constructed in order to underwrite "the robust certitude that permits the movement from practice to action without ever believing in the difference between construction, collection, immanence or transcendence."[19]

Once this intensive work has rendered laboratory substances into their ideal state, they can assume their role as representatives of nature in the theatre of experimental science. For this, it helps if they are represented by a "scientific" name and a symbol, and ultimately, a formula. The reform of chemical nomenclature proposed by Guyton de Morveau, Lavoisier, Berthollet and Fourcroy in 1787 aimed to provide a system in which, following Condillac's formula, "facts, words, and ideas would be but three imprints of the same seal." Thus, they banished any denomination evoking the geographical origin of the substance or the name of its inventor. The name was intended to reflect the composition of the named substance, thereby relegating the elements of its natural history to the level of irrelevant anecdote. The purified and stabilized laboratory substances, complete with labels specifying their composition, are now so many reliable reactants that will behave predictably, without any surprises. Thus, these substances can be compared to wild animals that have been domesticated and are then used as aids in exploring the untamed jungle. They constitute reliable points of reference, well-defined sites of scientific civilisation within the uncharted, promiscuous and prolific kingdoms of nature.

The nomenclature reform of 1787 promised much more than simply neutral terms for these ideally pure reactants. The names of the "simples" — the substances that could not be analysed any further — were considered to form the letters of a specifically chemical alphabet. Starting with these letters, the chemists could reverse the process of analysis by putting them together and thereby recompose the original compound substances. Thus, according to Lavoisier, the new nomenclature was "a faithful mirror" of both the facts and their associated ideas. This vision of the nomenclature as itself an analytical tool was based on the writings of the French philosopher, Étienne de Condillac, in particular, his book *La Logique*. Inspired by Locke's empiricist philosophy, Condillac privileged analysis as the "natural logic" of the human mind, which permitted the passage from the simple to the complex. For Condillac, and so for Lavoisier, this progression from simple to complex was the only method for understanding the world, as it necessarily paralleled the passage from the known to the unknown. Just as the child learned appropriately from nature by associating simple words with simple ideas born out of simple sensations, the apprentice chemist could only learn his art by starting with a knowledge of the properties of the simple substances. If the chemist sought to know the complex substances without knowing the simples of which they were composed he was, on the empiricist model, bound to fall into error. This philosophical approach lent itself to the presentation of chemistry as a "system of knowledge" that had particularly evident pedagogical consequences. Even though Lavoisier's text passes through the three kingdoms of nature in the standard order of mineral, vegetable, and animal; the material is now organized according to a strict sense of simple and compound matter. In addition, the substances are transcribed into a rational nomenclature intended to reflect their degree of complexity.[20] One crucial lesson of these philosophical developments is the importance of the arduous repetitive laboratory tasks of separation and purification in allowing chemistry to elaborate its own logic of "elements" and "compounds" that would in large part underwrite its claim to scientific autonomy.

A Social Space

The twentieth-century French philosopher, Gaston Bachelard, regularly emphasized the social dimension of science implied by the use of instruments. The deployment of scientific instruments requires not only the mobilization of a group of manufacturers and technicians, but also the establishment of conventional norms and standards, implying agreement across the whole experimental community. The chemists' labour of purification illustrates this mobilization of social resources necessary for the production (or construction) of a scientific fact so well that purification became Bachelard's favourite metaphor for describing what he referred to as "phenomenotechnics".

> In sum, one can say that there is no purity without purification. Nothing can better prove the eminently social nature of contemporary science than purification techniques. Indeed, the processes of purification can only develop by means of a whole range of reactants whose purity is underwritten by a sort of social guarantee. [...] All these theses would appear less superficial to the philosopher if he were only aware of the veritable factory processing (*usinage*) needed to produce a pure substance in contemporary technology. He would soon understand that this kind of purification is not an individual activity, that it demands a production line, a series of purifications, in sum, that the laboratory-factory has today become a fundamental reality.[21]

In historical terms, this laboratory-factory was first put into place in certain European universities in the nineteenth century where students were trained in the techniques of analysis and purification. The laboratory-school at Giessen set up by Justus von Liebig in 1830 was the first such enterprise, conceived with the aim of producing experienced chemists in as short a time as possible. Liebig submitted his students to an intensive practical training that could easily occupy them eight hours a day. A dozen or so students worked beside him in his own laboratory, with practice rendering the most complex and delicate operations into routine manipulations. This model of organization was disseminated throughout the industrializing world, being adopted by universities, colleges and schools in England, France and the United States.[22] The chemistry laboratory, whether institutional, domestic and personal, or industrial became a collective space populated by a hierarchy of personnel: heads of laboratories, researchers and technicians, all operating according to a rationalized and increasingly rigid system of division of labour. At the university or technical school, such laboratories served as an apprenticeship for the increasing numbers of students destined for careers in industry or the civil service.

While laboratories were becoming richer social spaces, they were also being transformed into official institutions supported by the state, by industry, or by wealthy patrons (often in the form of industrialists' foundations). Thus, chemistry was a key player in the global re-organization of knowledge in the West at the end of the nineteenth century. In this movement, science not only became an integral factor in industrial production, but also saw its practices transformed into social ones with the aim of disseminating technological efficiency, as well as obtaining economic mastery over the marketplace and reinforcing the military power of the nation-states.[23] On one hand, the laboratory-based schools supplied a scientifically trained workforce well versed in the most delicate techniques of analysis and purification. The success of these schools within and outside the academy contributed to the irruption of technology and industry into the traditional university system. On the other hand, the end of the nineteenth century saw the multiplication of research laboratories within industrial enterprises, particularly in the context of the synthetic dye industry, thereby underlining the double vocation of the laboratory as a space of both cognitive and

Figure 5. The "Piano of Chemicals": a collection of reagents for the identification of inorganic compounds. Used consecutively, the chemicals in the kit identify the presence of ions in solution. Photo by P. Cintra, courtesy of the Museum of Science of the University of Lisbon, Portugal.

technological endeavour. In this way, the laboratory was integrated into a range of institutions oriented towards the prize of technological innovation, and in turn wove the methods and values of industrial production into its constitutional scientific fabric.

An Instrumental Space

To conclude this chapter, we return to the furniture of the laboratory with which we started — experimental instruments. Any research or industrial chemistry laboratory contains a considerable amount of apparatus that permits rapid, simple, and relatively accurate analysis of substances in terms of their content. Traditionally, chemists determined the constitution of an unknown substance through its reactions with a number of well known reagents following standard procedures (see figure 5). Once again, Liebig is a pioneering figure in this domain. Thus, his *Kaliapparat*, which provided a relatively easy way to determine the carbon and hydrogen content of any given organic compound, was one of the attractions of his laboratory in Giessen, as it allowed his students to perform such analyses accurately and rapidly.[24] Precise and quantitative measures that can guarantee the purity of commercial goods, and can also be used to standardize them, not only served industry in the domain of quality control, but also contributed to the development of research into radicals in organic chemistry, and the applications of this research in chemistry, pharmacy and toxicology.

The multiplication of these analytical instruments served to give chemists social authority in their role as experts in legal proceedings (particularly in cases of suspected poisoning), and as experts in soil analysis, to cite but two examples. Indeed, during the nineteenth century, analytical chemistry became a speciality in its own right, implying not only the identification of a specific compound within a complex substance but also its isolation and purification. Whether the analysis employed gravimetric or volumetric methods, it necessarily mobilized chemical processes, and presupposed a thorough knowledge of the various reactants involved. The classic historical example involving the separation of an element from a mixture is Marie Curie's work to isolate radium from pitchblende using chemical separation techniques combined with physical measurements to monitor her progress. While such arduous repetitive work involving repeated crystallization made use of devices borrowed from physics to measure radioactivity or the optical properties of solutions, these instruments were secondary to the manual chemical manipulation of the material.

Today, things have changed, and modern chemistry laboratories are full of physical instruments, large and small, equipped with digital read-outs that relate the results of the analysis. With their pH meters, mass-spectrometers, nuclear-magnetic-resonance (NMR) apparatus, or infra-red spectrometers, research laboratories as well as industrial laboratories are filled with machines that measure, test, control and analyse (Figure 6). The deep transformation that took place between the 1950s and the 1970s has been described by Davis Baird, among other historians, as the second chemical revolution.[25] This change has deeply altered chemical practices. Instead of using chemical reactions to determine the constitution of substances, chemists now use a variety of instruments that allow non-destructive analysis of smaller samples in terms of their physical properties. Do these developments mean, however, as some critics have suggested, that chemistry has been colonized by physics?[26]

Figure 6.   This anonymous publicity photograph shows a woman using an infrared spectrometer (Perkin-Elmer Model 21). Not only is the instrument put in a clean uncluttered environment (except for the ashtray with a stubbed-out cigarette), but the framed photograph of the mushroom cloud from a nuclear explosion underlines the association that this type of instrument initially had with physics rather than chemistry. 1955, Applera/Perkin-Elmer Collection. Reproduced by courtesy of the Chemical Heritage Foundation.

Despite the predominance of these analytical instruments, which originated in physics laboratories, in today's chemistry laboratories, chemists have not become dependent on physicists. This is because the new instruments were not transferred as black boxes. The chemists who introduced NMR- or mass-spectrometers into chemistry laboratories interacted with the instrument-makers in order to adapt the physical instruments to the demands of their own field. Over time, instruments that were initially foreign to chemists became so familiar that they were able to provide theoretical interpretations of the data without the mediation of their colleagues from physics. Moreover, it was principally the chemists who disseminated the new techniques through their teaching and textbooks, thereby profoundly transforming the chemistry curricula. Thus, chemistry's ability to retain its methodological autonomy despite the introduction of these instruments may be due to the fact that they were not imported from physics along with established routine methods and applications, and had to be actively integrated. Indeed, it took several decades before these instruments were accepted as not only quicker and more efficient, but also as reliable enough to dispense with the complex and delicate manual analytical techniques that they replaced. Another result of this process of adaptation and appropriation has been that chemistry departments have come to require more and more large and costly pieces of equipment, making the discipline considerably more expensive. Furthermore, by adopting these high-tech instruments, chemistry not only entered the era of "big science", but also broadened the scope of chemical expertise, being able to analyse traces that are taken from ever smaller samples. In addition, the non-specific nature or polyvalence of these instruments can easily end up redefining research programmes in chemistry. Thus, while often introduced with the intention of investigating one class of substances, these analytical instruments can easily generate new fields of research or drive the same initial project on to cover new classes of substances.

This instrumental evolution has also brought about more subtle shifts in the culture of chemistry. The very conception of analysis has been changed by the use of these new instruments. Analysis is no longer a question of the "literal" division and subdivision of the original material to arrive at the constitutive simples as proposed by Lavoisier, but one of detecting properties of the (unanalysed) substance in question, by means of which the chemist can deduce its constituent parts. While the non-destructive nature of spectrometry offers one of its clearest advantages over traditional analytical techniques, the principle is a far cry from the experimental tradition of chemistry. The analytical techniques drawn from physics have also shifted the object of analysis. In the classical method using chemical reagents, the objects of the analysis were the elements understood in terms of their relations to other elements, and the disposition of the various components and compounds to react with one another. The interpretation of spectra — whether generated by absorption or emission — depends on the properties of atoms as interpreted by quantum mechanics. Furthermore, the visualization and manipulation of structures gradually became a major objective in itself. Thus, chemistry has found itself in a complex conceptual situation where individual atoms and molecules (and not manipulable macroscopic substances) have increasingly been pushed to the fore as the principle actors in chemical reactions.

Finally, the importance assumed by these analytical instruments has raised an important problem of qualifications in chemistry, in particular, the issue of how to differentiate between technicians and research scientists. This observation takes us back to our previous discussion of "tacit knowledge" as the foundation of experimental chemistry. The expert in instruments and making measurements possesses a know-how that is assessed in terms of efficacy, not in terms of incorporated skills, the chemist's *habitus* or intuition. A specialist in infra-red spectroscopy needs to be very familiar with the instrument, and competent in interpreting the results. He or she need not necessarily be an expert at carrying out the reactions that produce the substances under analysis. Thus, there is a form of "instrumental logic" that encourages the definition, redefinition, and even dissolution of disciplinary boundaries in contemporary research. While this is not unique to chemistry, it still poses a problem, as it does in other disciplines.[27]

In conclusion, then, while it has evolved a great deal over the past three hundred years, the laboratory has remained the privileged site for the elaboration of chemical knowledge, as it is the place where knowing confronts doing. Whether it is in the context of a secret, solitary quest for knowledge, or a social and hierarchized space of production, the laboratory has always been the place where chemists worked, ground up, transformed, assembled, or purified material substances in order to produce knowledge, often with the ultimate goal of producing commercial goods.

The final question with which we want to close this chapter is the following; where do chemists derive their confidence in the fruits of their laboratory work? How do they manage to convince themselves, and the wider public, that an artefact of their own construction effectively provides any information concerning natural products and processes? In other words, how can chemists escape the vicious circle denounced by Hannah Arendt in her book *The Human Condition*?

If, therefore, present-day science in its perplexity points to technical achievements to "prove" that we deal with an "authentic order" given in nature, it seems it has fallen into a vicious circle, which can be formulated as follows: scientists formulate their hypotheses to arrange their experiments and then use these experiments to verify their hypotheses; during this whole enterprise, they obviously deal with a hypothetical nature.

In other words, the world of the experiment seems always capable of becoming a man-made reality, and this, while it may increase man's power of making and acting, even of creating a world, far beyond what any previous age dared to imagine in dream and phantasy, unfortunately puts man back once more — and now even more forcefully — into the prison of his own mind, into the limitation of patterns he himself created.

Hannah Arendt, *The Human Condition*, Chicago: University of Chicago Press, 1958, pp. 287–288.

References

1. For, anthropological, sociological and epistemological considerations on the space and functioning of the laboratory, see B. Latour (1979) and H. Collins (1985).
2. N. Lemery (1675), pp. 375–376. English translation from Lemery (1677).
3. For the classic description of the disciplining of the human body through prescriptions and descriptions, see M. Foucault (1977). In Foucault's work, discipline has almost exclusively negative connotations, but thinking of the alchemical tradition we can give a positive, quasi-religious sense to discipline, understood as a prerequisite for attaining control over the material world and oneself.
4. Ursula Klein 'The Creative Power of Paper Tools in Early Nineteenth-Century Chemistry' in U. Klein ed (2001), pp. 13–34.
5. See F. Aftalion (1991), particularly Chapters 1–3.
6. This explains Holmes's ironic reflection: 'Why do historians continue to study the history of chemical theory instead of studying the history of their experimental practices?' F. L. Holmes (1989).
7. R. Siegfried (2002), p. 113.
8. A. Duncan (1996), p. 27.
9. J.-L. Thenard, *Essai de philosophie chimique*, in Thenard (1834–1836) vol. 5, pp. 409–519, citation on p. 409.
10. F. Bacon (1620) Part 1, aphorism 95, p. 349.

11. D. Diderot (1753), pp. 191–192.
12. J. Locke (1689), Book II, ch. IX, para. 8, p. 146.
13. The *habitus* describes the totality of someone's behaviour, both physical and mental, their way of being in the world. In sociology, this term is particularly associated with P. Bourdieu, notably P. Bourdieu (1979).
14. D. Diderot (1753), pp. 196–197.
15. See, J. Riskin (2002).
16. M. Polanyi (1958).
17. In light of its high cost, Diderot and d'Alembert's *Encyclopedia* was only destined for a wealthy clientele. It purported to describe all the arts and sciences of the day, although it could never live up to its claims for comprehensiveness. For a thorough history of the *Encyclopedia*, see R. Darnton (1979).
18. A.-L. Lavoisier (1789), translation by Robert Kerr.
19. B. Latour (1996), p. 44.
20. While this system worked well enough for the less complex inorganic compounds, Lavoisier was obliged to adopt the old-fashioned names for organic compounds meaning that the nomenclature system broke down in the vegetable and animal kingdoms. For a more detailed discussion of this issue, see J. Simon (2002).
21. G. Bachelard (1953), p. 78.
22. For a classic article on the importance of Liebig's laboratory teaching method, see J. Morrell (1972) and for a more recent study, see W. H. Brock (1997).
23. This transformation in the regimes of knowledge is described in D. Pestre (2003). For an illustration of this phenomenon in chemistry, we can look at H. Le Chatelier (1925) where the chemist argues for the reorganization of scientific research on the model of modern rationalized industry.
24. A. Rocke (2001), Chapters 1 and 2.
25. D. Baird (1993), and P. Morris (2002).
26. This issue is discussed in C. Reinhardt (2006).
27. As examples in other scientific domains, we can cite the importance assumed by computer programming in molecular biology and climatology, among others.

CHAPTER 5

PROOF IN THE LABORATORY

Earlier, we argued that the picture that Diderot and Venel presented of the chemist as a skilled artisan helped to construct a positive image of a laborious science in which all knowledge was earned at the price of the scientist's sweat and toil. Despite the various changes in chemical laboratories brought about by the introduction of physical instrumentation over the course of the twentieth century, being able to see at a glance — *le coup d'œil* — remains central to being a good experimentalist. This ability has oriented the science towards a particular method of investigation — the indexical method. Carlo Ginzburg has elaborated this kind of approach in his historical work, suggesting a parallel with Sherlock Holmes's method whereby one searches for any available clues and then follows the avenues that they open up.[1] This approach is as much about intuition as it is about any formalized or formalizable reasoning. However, neither the painstaking work of Diderot's heroic artisan nor the intuition of Ginzburg's detective-scientist are capable of establishing the chemical truth or of winning over other chemists to a proposed interpretation of nature. This is why the "sentimental empiricism" promoted by Venel and Diderot was replaced by Lavoisier's new experimental style of chemistry. In order to illustrate this transformation, and to try and understand what is involved in the "work of proof" — to use an expression of Bachelard's — we will examine one of Lavoisier's best known experiments involving the analysis and synthesis of water.

Chemical Experiment as Public Spectacle

On the 27 and 28 February 1785, Lavoisier invited his colleagues from the Royal Academy of Sciences, as well as certain courtiers and scientists visiting from abroad to attend his personal laboratory at the Arsenal in Paris.[2] Thus, a select group of about thirty hand-picked guests came to witness a chemical experiment that would become, if it was not already, emblematic of the chemical revolution. What Lavoisier proposed to do in the presence of the assembled dignitaries was to decompose a significant volume of one of the Aristotelian elements — water — and to collect and weigh the products of this decomposition. As if this were not spectacular enough, he would then recombine these new elements into water. All this in a display surrounded with measures of purity and barimetric verifications to dispel any lingering suspicions of trickery or sleight of hand. Thus, this public experiment can be seen as a further extension of several key developments in modern experimental science. Like Boyle with his air pump, Lavoisier needed to strike a balance between the spectacular and the convincing. What was presented was intended to run counter to the deepest convictions of an Aristotelian audience (who believed that water was an unanalysable element) and yet be above any suspicion of trickery or the showmanship with which these cosmopolitan figures would have been only too familiar. The use of transparent apparatus, in particular glass reaction vessels, was a technological key, albeit one among many, for demonstrating the invisible operations of nature to the public. Thus, while one cannot "see" a vacuum or the synthesis of water, one can see that nothing unnatural or untoward is going on under the cover of the opaque wall of a metal jar or some other reaction vessel.[3]

Lavoisier was also borrowing from a more recent style of *mise-en-scene* of the spectacle of science with Abbé Nollet as its leading proponent. In Nollet's shows, the protagonists were the newly domesticated high-voltage electrical phenomena, providing demonstrations that amazed and shocked a bourgeois Parisian public eager to witness such scientific novelties. Although the water experiments were intended to be spectacular, Lavoisier was not working in the same register as the detonations and sparks of the electrical display. Instead he chose a sober style, more suitable for a leading figure of the Royal Academy of Sciences and in line with his goal of the experimental revelation of a profound truth of nature.

This kind of chemical demonstration, which combined the carefully controlled analysis and synthesis of a substance, would be explicitly held up in the nineteenth century as a model for other sciences to imitate. Thus, psychologists and historians, among others, mobilized just this model in their quest to attain the coveted status of a science.[4]

Here, however, we want to examine where exactly the probitive power of this public experiment lay. What needed to be done, and what suppositions needed to be made in order to raise the observation of a few drops of humidity on the inside of a glass vessel to the status of scientific proof? Indeeed, Lavoisier was setting himself up against the longstanding popular and scientific opinion that water was an unanalysable element. This was not entirely new territory for him, though, as in 1777 he had already worked on the decomposition of another element — air. To do this, he had exploited a number of discoveries that had recently enabled chemists to differentiate between three different "distinct" types of air: fixed, dephlogisticated, and inflammable air. Lavoisier took them to be separable components of "common air" (atmospheric air) rather than various forms of elementary air.[5] In July 1778, Lavoisier read a memoir to the Academy of Sciences, entitled "On the combination of the matter of fire with evaporable fluids and the formation of elastic aeriform fluids". In this memoir, he argued that all the airs or aeriform fluids were composed of a characteristic base or radical combined with the matter of fire, a substance he would later term caloric.[6] Indeed, Lavoisier thought that the combination with caloric explained the gaseous nature of certain substances, while the "base" bore the other, characteristic properties of a gas like oxygen or hydrogen (his names for dephlogisticated and inflammable air, respectively). In principle, if this "base" could somehow be separated from its caloric, one could obtain the substance in a liquid or solid form. Therefore, while the different gases might be similar in appearance, they were fundamentally distinct. Lavoisier's physiological experiments seemed to confirm this opinion, as the lungs of living organisms fixed only a certain part of what they inhaled. Thus, the animal metabolism served to analyse atmospheric air in the same way as combustion or calcination, converting dephlogisticated or vital air (oxygen) into fixed air (carbon dioxide).[7]

The decomposition of water did not, however, follow in any logical sense from the decomposition of air. While several chemists, including Lavoisier, had tried burning hydrogen in air or oxygen, often with the idea that they would obtain an acid or even fixed air, they seemed to obtain nothing at all.

Two events appear to have triggered Lavoisier's experiments on the decomposition and recomposition of water. First, reports reached Lavoisier's ears in 1783 that the English chemist Henry Cavendish had produced water by igniting a mixture of inflammable and vital air. Second, in July 1783, Lavoisier was appointed, along with Monge, to a five-man commission of the Royal Academy of Sciences to review Montgolfier's invention of the hot-air balloon. Their study of the functioning and potential of these balloons led the commission to reflect on alternatives to hot air for filling them. Inflammable air (or hydrogen) was an interesting candidate due to its low density and the availability of easy methods for producing it.[8]

Instruments of Decision

Let us now turn to a more detailed examination of this double experiment involving the analysis and synthesis of water. Consider first the result of the second part, the recomposition of water from its constitutive gaseous elements. What did Lavoisier observe on the walls of the reaction vessel, before going on to collect and carefully weight it? It was water, water in its abstract modern chemical sense, not water from Evian or Paris. Water possessing its "characteristic properties" as a chemical substance, but devoid of any more individual ones. Thus, this water participated in this general movement of abstraction associated with the laboratory that implies stripping away any variability and singularity associated with the material's natural occurrence.

How did this process of abstraction or universalization work in this case? First, it required fixing the identity of the reagents, inflammable air (hydrogen) and vital air (oxygen). At the time of this experiment, however, identifying these gases was not an easy matter. There was a debate between chemists as to whether the hydrogen obtained from the reduction of water was the same as that derived from marsh gas or from some other source. Furthermore, Lavoisier needed to convince his audience of the nature of the final product, that it was indeed water. Lavoisier was not in a hopeless situation, however, as he had inherited a repertoire of tests from the generations of chemists that had preceded him, including the use of lime water to identify fixed air, as well as flame and animal tests that could distinguish between oxygen and hydrogen, for example.[9] He also disposed of various indicators that had been developed starting in the 17th century, including the Litmus test which was extensively used by Boyle.

On a visit to Paris in the spring of 1783, Charles Blagden told Lavoisier how Cavendish had obtained pure water by igniting a mixture of inflammable air and dephlogisticated air in a closed vessel. Thus, the synthesis of water was already known among a group of chemists when Lavoisier performed his own experiments, but the phenomenon still needed to be verified. This was even more important for Lavoisier, because, unlike Cavendish, he wanted to argue that it involved the constitution of water from its elements and not simply the transformation of the element from one form into another. How could Lavoisier persuade his contemporaries, most of whom were convinced that water was an element, that it was in reality a compound of two other elements? It was clear that he would have to pull out all the stops, and he would indeed spare no expense in his efforts at persuasion, as we shall see. He needed to undermine some of his colleagues' most deeply held beliefs, then exploit the resulting instability, making the balance of their opinion shift from one certainty via an intermediary, finely balanced stage of uncertainty, before nudging them towards a firm new conviction. The analogy with the operation of a pair of scales is deliberate, as this was the instrument that the chemist would privilege in his efforts to change the others' minds about the nature of water.

Over time, the pair of scales has become an emblem for Lavoisier as the founder of modern chemistry. The scales were the instrument used to operationalize his foundational maxim that during a chemical reaction: "nothing is created; nothing is destroyed". While this concept of the conservation of mass promised to displace chemistry from the largely qualitative "alchemical" ontology of sympathies, antipathies, and ethereal principles, it also served as a powerful metaphor for the process of rational thought; the key to the persuasive power of experiment. Lavoisier's gravimetric experiments were intended to tip the balance in favour of his interpretation of the nature of water!

Nevertheless, we want to insist on the fact that Lavoisier did not invent either the idea of the conservation of mass in a chemical reaction or the use of the scales in chemistry. The idea that nothing can be created or destroyed can be found in the writings of the ancients, and many physicists and chemists had championed it as an axiom before Lavoisier. Van Helmont, for example, explicitly proposed that: "Nothing comes into being from nothing. The weight comes from another body weighing just as much."[10] Despite what a number of historians or chemists have maintained, Lavoisier did not introduce the precision scales into chemistry either, as they had been used by alchemists, apothecaries, and assayers for centuries. They were essential to a number of qualitative tests for purity, and were indispensable for determining the specific weight of metals, a key to ensuring the authenticity of coins or establishing the nature of amalgams of all sorts.

Nevertheless, Lavoisier's scales were more than a precision instrument. They materialized an intellectual strategy of balancing inputs and outputs that Lavoisier used daily in his book-keeping activity as a tax collector and also in his reflections on rational economics — both domestic, and national. Lavoisier generalized the principle of weighing the pros and cons, deploying this way of assessing the worth of an argument in all the various domains where he was active throughout his life.[11]

While we are considering the question of Lavoisier's originality, we can also ask whether he was the first to conceive this form of paired reactions exemplified by his decomposition and recomposition of water, that is, experiments involving the analysis and synthesis of a substance as the means of determining, or persuading others of its true composition. It seems that this too was not original, as alchemists had long practiced the complementary processes of destruction and reconstitution as a way to counter accusations of trickery. Indeed, certain historians have argued that the term "spagyric" (a word associated with alchemy) might have originated from a combination of the Greek words *span* (to separate) and *agerein* (to put together).[12] Paracelsus placed analysis at the centre of his chemical practice, but, as he was more concerned with preparing medicines than with demonstrating the composition of bodies, he concentrated on separation (*Scheidung*) at the expense of recomposition. Nevertheless, at the beginning of the seventeenth century, both van Helmont and Daniel Sennert often paired analysis with synthesis to prove their chemical arguments. Furthermore, historians who have looked closely at van Helmont's experiments — in particular, the decomposition and recomposition of glass — have shown the importance of precise determinations of the weights of reactants in his attacks on Aristotelian principles.[13]

These "chymists", therefore, clearly accepted that decomposing and reconstituting compounds was a legitimate route to the truth. They believed that the thing in itself — an element that enters into the composition constituting the mixt, for example — could be shown to exist by the operations that make it disappear and then return it to existence. It is precisely this aspect of the chemical approach that provoked Kant's admiration of Georg Ernst Stahl's theory and led Kant to regard him as one of the founding figures responsible for putting physics on "the great route of science".

> When Galileo caused balls, the weights of which he had himself previously determined, to roll down an inclined plane; when Torricelli made the air carry a weight which he had calculated beforehand to be equal to that of a definite volume of water; or in more recent times, when Stahl changed metals into oxides, and oxides back into metal, by withdrawing something and then restoring it, a light broke upon all students of nature.[14]

Indeed, the demonstration by analysis and synthesis was constitutive of chemistry, and had been tested and perfected over the course of centuries, granting the chemist a means to illuminate the nature of a substance or class of substances. Chemists forged this powerful method of demonstration in the context of controversies. Such controversies are necessarily theoretical as they concern the principles taken to compose nature itself. That such debates are theoretical does not, however, mean that they are purely speculative. They have a very practical sense in so far as they put the very status of the chemist into play. Throughout the history of the science, the chemist has regularly been suspected of being a charlatan, and has, as a corollary, always been very concerned that he should be worthy of consideration as a natural philosopher on the same basis as the experimental or theoretical physicist.

The Power of Scepticism

This elaborate procedure for enacting an experimental demonstration responds to a profound criticism that had been the bane of chemists at least since the time of van Helmont. How can you be sure that the principles you have revealed by means of analysis were present in the compound from which you obtained them? How can you rule out the possibility that these substances were not produced *de novo* by the chemical operations employed in the analysis? Indeed, Boyle had used this very argument in his *Sceptical Chymist* to cast doubt on the existence of the various principles proposed by his contemporaries. No matter the heights of their skills in the art of decomposing matter, chemists could never escape the suspicion that what they considered to be elementary constituents of the substance at hand were in reality nothing more than experimental artefacts. One way of addressing this criticism was to reduce the intensity of the heat used to drive reactions, favouring the use of the mediated heat of a water bath or long macerations in water. In order, therefore, to demonstrate the composition of a substance, the analysis needs to be paired with a "syncrese", a synthesis or recomposition of the initial substance.

Can this kind of complementary synthesis serve to allay all the doubts of the sceptical chemist? If we look more closely at the case of Lavoisier's analysis and synthesis of water, the answer seems to be no. Having heard of Cavendish's experiment in June 1783, Lavoisier worked with Laplace to demonstrate the composition of water by synthesis. Lavoisier used his "pneumatic receptacles" to supply known volumes of the two gases to a common glass vessel where he ignited the mixture using an electric spark. Using a process of trial and error, they determined the correct proportion of the two gases, considering it to be the one that gave "the most luminous and beautiful flame". Upon ignition, they saw the glass reaction vessel become cloudy with vapour that condensed into droplets covering the interior within about twenty minutes. They used a series of tests to ensure that this was water — sunflower water, tincture of violets, lime water, and others. The next day, they reported to their colleagues at the Academy of Sciences that "water is not a simple substance, and is composed pound for pound of inflammable and vital air."[15] The minutes of the experiment, signed by Blagden, Séjour, Laplace, Lavoisier, Vandermonde, Fourcroy, Legendre, and Meusnier, testify to the validity of the conclusions read at a public meeting of the Academy.[16]

This synthesis was not, however, conclusive, even for those who performed it. As the historian Maurice Daumas has pointed out, assertions about the purity of the water relied on purely qualitative results.[17] Lavoisier used an axiom drawn from geometry to make up for the lack of quantitative data in this area.

> Because the two airs were brought from the pneumatic receptacles to the jar by means of flexible pipes made out of leather, and these were not absolutely impermeable to the air, it was not possible to be sure of the exact quantity of the two airs when we carried out the combustion. Nevertheless, as it is no less true in physics than in geometry that the sum is equal to its parts, and as we obtained only pure water in this experiment, without any other residue, we believe that we are entitled to conclude that the weight of this water is equal to that of the two airs that served to form it.[18]

Judging by a letter that he wrote to a colleague, Jean Antoine Deluc, a few days later, Laplace was very sceptical about this experiment. Lavoisier too was conscious of the weaknesses in this demonstration and on the occasion of the publication of a memoir in the *Histoire et mémoires de l'Académie,* he added the quantitative results obtained by his colleague, the mathematician Gaspard Monge. Monge had also performed the synthesis of water at the Mézières Military Academy using very precise measurements not only of the volumes of gases, but also of the weights of all the substances involved. While useful, the cumulation of concordant data was insufficient to prove the composition of water. That is why Lavoisier performed the experiment again in 1784, and then prepared his grand formal demonstration for February 1785.

The Price of Proof

Considering this experiment of the decomposition and recomposition of water leads us to a third general remark, that this process of analysis and synthesis is not as straightforward as it may appear. The proof is made to seem easier than the experimental reality, which required the mobilization of all sorts of resources, theoretical, technical, and straightforwardly practical. Thus, if we consider just the process of analysis, Lavoisier had to enlist the aid of Jean-Baptiste Meusnier, a former student at the Military Academy in Mézières. The first question was whether to attack the hydrogen or the oxygen contained in the water to liberate the other component.

To answer this question, Lavoisier used the wealth of empirical knowledge contained in the affinity tables that he had studied in his chemistry course with Rouelle (see Figure 4). He knew that in order to displace a substance B from a combination AB, one needed to introduce a substance C that had a greater affinity than B for A. This reasoning led him to use the marked affinity of iron for oxygen to remove the "oxygen principle" from the water. Deploying Meusnier's skills as an engineer, Lavoisier perfected an experiment involving a heated tube of iron — the barrel of a gun — into which he introduced the water, one drop at a time.

> The water completely decomposes, and no part of it exits by the inferior opening of the barrel; the oxygen principle of the water combines with the iron, calcining it. At the same time, the freed acqueous inflammable principle passes into the aeriform state, with a specific gravity about two twenty-fifths of common air.[19]

The experiment appears straightforward enough. Nevertheless, there were plenty of technical problems. The barrel itself tended to decompose, and to avoid this, it had to be wrapped in copper wire. Furthermore, it was far from obvious how to measure the increase in weight of the iron due to the process of calcination, a determination that was needed to check the amount of oxygen principle released by the decomposition of water.

The kind of quantitative demonstrations that Lavoisier was looking for required much more sophisticated apparatus than the pneumatic trough that had served Priestley and Cavendish so well. Again in collaboration with Meusnier, Lavoisier had a new type of pneumatic receptacle custom built by Mégnié and Fortin, the finest instrument-makers in Paris. These imposing gazometers consisted of large brass containers with pistons (Figure 7) that used a weight-operated cantilever system to keep the gas at a constant pressure even while it was being used up in a reaction. This had the double advantage of providing a constant flow of the gas to the reaction vessel and measuring the volume being used, from which Lavoisier could easily calculate the all important weight of the reactant.

After investing a considerable amount of time and money in this sophisticated and very expensive apparatus for handling the gases, Lavoisier and Meusnier decided to profit from this investment. They would put the

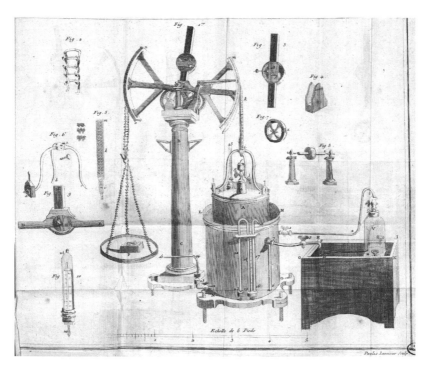

Figure 7. Lavoisier's gazometer designed by the engineer Jean-Baptiste Meusnier and drawn by Marie-Anne Paulze Lavoisier. It allowed the measurement of the weight of a gas consumed in a reaction. This large, expensive piece of apparatus performed an essential role in the experimental demonstration of the synthesis of water. Plate from Lavoisier's *Elements of Chemistry*, 1789, image courtesy of the Service Interétablissements de Coopération Documentaire des universités de Strasbourg Département du Patrimoine. http://num-sed-ulp. ustrasbg.fr:8080/.

spectacular decomposition and recomposition of water on show before Paris's leading scientific figures to win these opinion makers over to their interpretation of the phenomena. The preparations for these public experiments took two months, keeping Lavoisier and Meusnier busy from 21 December until 12 February, calibrating instruments and fine tuning the experimental set-up for the most impressive part, the synthesis of water. They also had to collect enough pure gaseous reactants, as well as taking care of other more or less important details that would ensure the smooth running of the public demonstration.

When the public experiment finally took place, it lasted two full days. Lavoisier started at 11.59 on 27 February with the decomposition of water, and only finished at 18.30 with the careful weighing of the hydrogen, which had been collected in ten large glass jars. Meanwhile, he made his first attempt to synthesize water in the afternoon on the same day. This involved filling a gazometer with oxygen and another with hydrogen, taking every precaution to keep air out of the containers. Tubes connected these two gazometers to the reaction vessel in which the mixture of gases was ignited using an electric spark generated by one of Lavoisier's electrical machines based on a design by Ramsden. They succeeded in igniting the mixture on the third attempt, and then kept the reaction going for 10 hours, all the time carefully measuring the consumption of hydrogen and oxygen. During the night, Lavoisier had the hydrogen generated in the decomposition reaction transferred into the gazometer providing raw material for the recomposition. At eight o'clock the next morning, the assembled scientists verified the levels in the gazometers. At 11.39, the decomposition reaction was restarted, followed by the synthesis. The experiments were wound up at 23.35 on 28 February, when Monge placed official seals on the reaction vessels for the synthesis, thereby marking the end of the public presentation, but not the demonstration. Several more days were needed to conclude the experiments by establishing the precise weights of the reactants involved. The next day the quantities of reactants were established in the presence of four official witnesses from the Royal Academy. For the synthesis, the gas remaining in the gazometers, the water in the reaction vessels (5 ounces, 4 gros, 54.5 grains, or about 175 grams) as well as the caustic potash used to purify the gases were all weighed. Qualitative tests were also performed on the "water" that was formed. A week later, on 7 March, the gases used for the synthesis were analysed, particularly the oxygen, which had been obtained from mercury oxide. The instruments and reaction vessels were also examined and weighed, and the water obtained in the reaction vessel was tested once again. Thus, it was only on 12 March that the members of the commission could meet to sign off on the official gravimetric results of the experiment.

Two weeks to carry out a pair of complementary reactions and this does not even take into account the months of preparation. Today, this kind of experiment would be unthinkable outside the context of public-funded or industrial scientific research. Lavoisier, however, had paid for it all — 1 814 livres (about £36 000) — out of his own pocket. However high, it was a price Lavoisier was prepared to pay to establish the truth.

The Limits of Proof

We have tried our best to convey the persuasive power of Lavoisier's elaborately staged experiment, but the question remains; did it really prove anything? One could object that the reactants used to synthesize the water were not actually derived from the decomposition experiments. Lavoisier's unrelenting pursuit of experimental gravimetric precision might be seen to be an attempt to cover up for this kind of logical flaw in the demonstration of analysis and synthesis. Nevertheless, on closer inspection, his obsession with precision was not so much an inflexible rule governing every detail of his experiments as a way of conceiving this practice. On the second day, a vessel accidentally broke, splashing some of the all-important synthesized water onto the wall. This did not, however, invalidate the experiment in Lavoisier's eyes. Thanks to some judicious trials in which different amounts of water were splashed onto the wall, Lavoisier was able to come up with a precise value for the weight of the water that had been lost in the accident.

More striking, however, is the reaction of the members of the audience who witnessed the experiment. While Berthollet was won over to Lavoisier's side, other members of the Academy like Baumé, Cadet and Sage remained convinced that water was an element, while Fougeroux de Bondaroy was simply confused by the whole affair. Berthollet proved a valuable ally and proselyte for Lavoisier, but this experiment did not prevent a number of chemists from remaining sceptical concerning his views. Thus, while Lavoisier succeeded in converting the Dutch chemist Martinus van Marum when he came to visit his laboratory on a trip to Paris in the summer of 1785, others, including Senebier, Fontana, Priestley, and even Cavendish, were never won over. Indeed, the principal result of this "demonstration" seems to have been a heightening of the debate between Lavoisier's supporters and his opponents.[20]

It would, however, take more than Lavoisier's demonstration of the synthesis of water to make the composite nature of this Aristotelian element a stable and robust fact, no matter how compelling the spectacle may have been for some of his assembled guests. Indeed, Lavoisier would repeat this manipulation himself in a number of versions, including a series of experiments conducted in collaboration with Laplace in 1786–1787 to determine the amount of heat generated by the combustion of hydrogen. Nevertheless, many of those who were unreceptive to Lavoisier's interpretation remained unconvinced, even when they succeeded in conducting similar experiments themselves. Priestley never changed his mind on the question, and he was not alone. The adoption of Lavoisier's views on the composition of water was a complex process heavily dependent on the local cultures of chemistry, including the political and philosophical commitments of those concerned.[21]

Other chemists, both inside and outside France, devised less costly means to repeat Lavoisier's experiments with hydrogen and oxygen, and were able to arrive at quantitative results for the composition of water.[22] Replication was, as in other cases in the history of science, necessary if not sufficient for Lavoisier to convince his peers at once of the reality of the phenomena and the innovative interpretation he wanted to give them.

There are three traits of chemical proof that are illustrated by the exemplary case of the decomposition and reconstitution of water. First, chemists materialize the abstract processes of analysis and synthesis in terms of chemical operations and observable phenomena, an approach that distinguishes chemistry from geometry, for example, where analysis and synthesis were equally prominent processes. We can describe this feature of chemistry using the term "rational materialism" deployed by Gaston Bachelard, even though he himself, under the influence of Carl Jung's interpretation of alchemy as a spiritual quest articulated around images and symbols rather than as a laboratory practice, refused to apply this term to alchemy.[23] Nevertheless, it is clear that the seventeenth and eighteenth-century chemists who had dealt with decomposition and reconstitution were the pioneers of a science that considers practical manipulations as the ultimate proof of veracity.

A second characteristic feature of this art of proof — this time unfalteringly identified by Bachelard — is that while it may be experimental, this does not mean that it cannot be theoretical in its implications. Just because chemists work with their hands does not mean that they do not work with their minds at the same time. Every step of Lavoisier's demonstration is loaded with theory. This is not only because it touched on fundamental principles such as the conservation of matter, but also because the reasoning behind many of the individual manipulations mobilized or challenged established theories. Once again, we can reaffirm the philo-sophical position that there is no such thing as theory-independent facts, and from this perspective it is unsurprising that chemists, like physicists, cannot prove their theoretical positions by means of bare empirical facts. Even the compilations of chemical "recipes" for pharmaceutical prepara-tions like those found in Lemery's treatise, which reproduced empirical knowledge collected over centuries of traditional craft labour, were impregnated with theory. While this theory is restricted to a community sharing a common scientific culture, there is no need to make this theory explicit. Indeed, it seems quite appropriate to use Kuhn's notion of "para-digm" in this context, as scientists who share a common paradigm do not need explicitly to evoke its principles or postulates, which form a shared set of references for all those concerned.

Third, the demonstration by analysis and synthesis not only mobilized theory, but also abstraction, another capacity often denied to chemists. Indeed, to have the demonstration function, Lavoisier needed to insist on the purity of the raw materials he used, as well as the abstract, universal nature of the products. The natural history of these elements and compounds was deemed irrelevant, and they were taken to have no specificity, with their material nature limited to the properties that characterized them as partici-pants in the reaction. In effect, Lavoisier used them as hypostatic principles, materializing his geometry-style combinatorial chemical reasoning. Thus, in a paradoxical move, this very materialization of Lavoisier's chemistry in the demonstration of the synthesis and analysis of water involved a complemen-tary idealization of the material bodies that he put into play.

One last point for reflection: what were the grounds for thinking that the analysis and synthesis reactions presented by Lavoisier really were com-plementary, symmetric reactions? Once again, Bachelard clearly identified this problem, and placed it at the centre of his book entitled *Rational Materialism*.

When one is led to reflect upon the relationship between synthesis and analysis, one is too often satisfied with identifying it as a dialectic of reunion and separation. This, however, is at the cost of an important nuance. Indeed, in modern chemistry, synthesis is the very process of invention, a process of rational creativity in which the rational plan for making an unknown substance is posed from the beginning as the problem that leads to the project. We can say that synthesis represents a process of penetration for modern chemistry, progressively penetrating in the course of realizing the project.[24]

The philosophical significance of synthesis is precisely the issue that we will address in the next chapter.

References

1. C. Ginzburg (1989).
2. For a detailed account of this experiment, see M. Daumas & D. Duveen (1959).
3. For more on the rise of experimental philosophy in the seventeenth century, and Robert Boyle in particular, see S. Shapin & S. Schaffer (1985).
4. To see this use of chemistry as a model for history, for example, N. Fustel de Coulanges (1888) and C.-V. Langlois and C. Seignobos (1898).
5. The first such air was fixed air (carbon dioxide), that J. Black (1754) liberated from magnesium carbonate, and identified using lime water. Henry Cavendish produced inflammable air (hydrogen) in 1766, and dephlogisticated air or vital air (oxygen) was identified by Priestley and Scheele in 1774.
6. A.-L. Lavoisier (1862–1896), vol. II, pp. 212–224.
7. See 'Expériences sur la respiration des animaux et sur les changements qui arrivent à l'air en passant par le poumon,' A.-L. Lavoisier (1862–1896), vol. II, pp. 184–193.
8. The potential military uses of these balloons encouraged investigations into producing hydrogen in large quantities, generating techniques that would facilitate Lavoisier's water experiments. We would argue that the intimate connection between such technical applications and 'fundamental' research is the norm rather than the exception in chemistry.
9. These primitive bioassays involved placing animals in an atmosphere of the gas under test. All the gases except oxygen would kill the animal.
10. V. Helmont, Ortus *'Progymnasma meteori'*, No 18, p. 71, quoted in W. Newman and L. Principe (2003), p. 69.
11. B. Bensaude-Vincent (1992).

12. This proposed etymology is discussed in W. Newman and L. Principe (2003), p. 90.

13. See, W. Newman and L. Principe (2003), pp. 62–66, and W. Newman 'Alchemy, Assaying, and Experiments,' in F. L. Holmes and T. Levere Eds (2000), pp. 35–54.

14. I. Kant (1963), p. 20.

15. A.-L. Lavoisier and J.-A. Chaptal, 'Mémoire dans lequel on a pour objet de prouver que l'eau n'est point une substance simple, un élément proprement dit, mais qu'elle est susceptible de décomposition et de recomposition', *Histoire et mémoires de l'Académie royale des sciences de Paris*, 1781 (1784), pp. 468–474, in A.-L. Lavoisier (1862–1896), vol. II, pp. 334–359.

16. They were later published as the memoir cited in the preceding footnote.

17. A.-L. Lavoisier, *Registres de laboratoire, archives de l'Académie des sciences*, vol. 8, p. 63. M. Daumas (1955), p. 143.

18. A.-L. Lavoisier (1862–1896), vol. II, pp. 338–359.

19. *Ibid*, p. 373.

20. J.-C. La Métherie, the editor of *Observations sur la physique* published a review of this experiment at the beginning of 1786 in which he opposed Lavoisier's interpretation of the phenomena. See, J.-C. La Métherie (1786). There is another critical note in this same volume on page 315.

21. J. Golinski, (1992) and J. McEvoy 'Language, liberty and chemistry in the English Enlightenment' in B. Bensaude-Vincent, and F. Abbri eds (1995).

22. Van Marum was the first to repeat the experiments on the decomposition and recomposition of water, performing them in Leyden in 1787. Louis Lefèvre de Gineau, professor of physics at the Collège Royal in Paris, also presented the experiments in public between 27 May and 7 June 1788. Like Lavoisier, he made very precise weight measurements, although he allowed a certain margin of error. De Gineau produced more than a kilogram of water and this large-scale experiment had a considerable impact on French chemists. In 1789, Von Troostwijk, Deiman and Cuthberson carried out another decomposition–recomposition experiment in Holland, which Sylvester and abbé Chappe tried to reproduce in Paris. Fourcroy, Armand Seguin and Nicolas Vauquelin then carried out the synthesis once again in the presence of commissioners from the Academy of Sciences. Based on this last experiment, the chemists concluded that

oxygen constituted 85.7% and hydrogen 14.3% by weight of the synthesized water.

23. Gaston Bachelard emphasized the rupture between alchemy and the modern science of chemistry, which 'follows a path of progressive clarification, while the alchemist waited to be illuminated', G. Bachelard (1953), p. 26.

24. *Ibid*, p. 23.

CHAPTER 6

CHEMISTRY CREATES ITS OBJECT

"Chemistry creates its object. This creative faculty, akin to that of art, forms an essential distinction between chemistry and the other natural or historical sciences."[1] This famous claim made in 1876 by the French chemist Marcellin Berthelot has been regularly cited by chemists throughout the 20th century, notably by two Nobel laureates: Robert Burns Woodward in 1956, and Jean-Marie Lehn in 1987.[2] This phrase has continued to ring true for chemists for more than a century, despite the profound transformations undergone by the science. The aim of this chapter is, therefore, to try to understand what essential truth about modern chemistry is conveyed by the claim that chemistry creates its object, at least in the eyes of the chemists themselves.

At first sight, it seems that Berthelot was just trying to distinguish between the experimental and the observational sciences, like natural history or astronomy, with the former constructing its objects in the laboratory and the latter simply observing them as they occur in nature, or at most isolating them to facilitate the task. When Berthelot used the term "create," however, he wanted to stress other connotations of the word, as we can see in the rest of the passage that follows the famous citation:

> The object [for the observational sciences] is given in advance and is independent of the scientist's will and activity: the general relationships that they can discern or establish are based on more or less probable inductions, or even on simple conjectures that it is impossible to verify beyond the external sphere of observed phenomena. These sciences do not possess their object. Thus, they are too often condemned to eternal impotence in their search for the truth, or have to be satisfied with possessing only a few sparse and often uncertain fragments.[3]

Indeed, the terms "will", "possess" and "impotence" all suggest the power of the chemist in the guise of a demiurge. But it would be a mistake to think that when he wrote these lines, Berthelot had in mind a notion of material power, rather, he was thinking of a form of power derived from knowledge. Berthelot did not have our contemporary vision of chemical synthesis as a means for generating the millions of new compounds that characterize our modern world. For him, synthesis was a means for arriving at a better understanding of nature, with as its unique goal the simple imitation of natural substances. We can see this in the conclusion to his book.

> Chemistry possesses this creative faculty to a higher degree than all the other sciences because it penetrates deeper, attaining the natural elements of things. Not only can it create phenomena, but it also has the power to form a multitude of artificial entities similar to natural ones, and sharing all their properties. These artificial entities are the instantiated images of abstract laws, that [chemistry] seeks to know [...] Without leaving the sphere of legitimate ambition, we can hope to conceive the general types for all possible substances and create them; we can, I claim, hope to recreate all the substances that have been developed since the very beginning and form them in the same conditions, according to the same laws, using the same forces that nature put into action to do so.[4]

Berthelot conceived of synthesis, like analysis, as a tool for knowing the world, a means for penetrating the secrets of the composition of natural substances. The fact that synthesis can redo what analysis has undone led him to the following definition of synthesis as a cognitive tool.

> The reproduction of the complete set of natural compounds using the elements in partnership with only the play of molecular forces and the chemical metamorphoses that matter undergoes in living beings.[5]

Thus, Berthelot's formulation "chemistry creates its object" was also a response to the idea that "chemistry destroys its object", a criticism often levelled at the practice of analysis by distillation due to the lack of resemblance between the original analysed substance and its products. The question remains, however, as to how synthesis can constitute a cognitive tool.

The Different Meanings of Synthesis

In common usage, we treat the adjectives artificial and synthetic as inter-changeable. But synthetic has a variety of meanings that are betrayed by this conflation with artificial. Synthesis implies the bringing together of different components to form a union, while artificial is defined in opposition to natural and does not suggest any similar process of combination. In chemistry, the composition of a whole by combining several distinct elements covers a number of different practices. While it is true that we refer to a substance not found in nature as synthetic, we also talk of synthetic when we are dealing with a copy of a naturally occurring compound. Thus, the salicylic acid found in aspirin tablets is the product of a partial synthesis of the pharmaceutical agent formerly extracted from willow bark. Although rarely done, it is also possible to synthesize organic compounds starting from their constitutive elements, carbon, hydrogen, oxygen, etc., a process that Berthelot termed "total synthesis".

Berthelot claimed that his objective was "to use the elements in order to construct the immediate principles of material entities by means of art".[6] Thus, he was not aiming to carry out the total synthesis of either plants or animals starting from carbon, hydrogen, oxygen, nitrogen, phosphorus and sulphur, but only aimed to synthesize the immediate principles, in turn obtained from the partial analysis of organic material. In contrast to elementary analysis that resolves a body into its constituent elements to determine its ultimate composition and their relevant proportions, this partial analysis is a "gentle" decomposition that does not destroy the nature of the chemical substances. Historically, this partial analysis turned around solvent extraction, with many key solvents only introduced in the course of the seventeenth and eighteenth centuries.

> Instead of destroying the organic substances right away and rendering them into elementary bodies, one carries out a gradual reduction, transforming these substances into simpler compounds, and so by degrees, one renders the complex and mobile principles formed under the influence of life down to artificial, simpler and more stable principles. The latter principles in turn become the object of new analyses of the same type, generating simpler and more stable principles, and so on, until one arrives at the elementary bodies.[7]

As far as the compounds drawn from the mineral kingdom are concerned, synthesis and analysis often prove to be complementary reactions that the chemist can carry out. Oxides, for example, can be converted into metals and the metals reconverted into oxides with relative ease. As we saw with the synthesis of water, once the elements that enter into a compound are obtained it is usually possible to find the means to recombine them to produce the original compound. While the requisite conditions may be difficult to realize, the successful analysis of a compound sets the chemist on the right path for the synthesis. This is not the case, however, for compounds derived from the animal and vegetable kingdoms. Organic compounds are composed of a limited number of elements combined in varying proportions, and their elementary analysis provides few clues for the means to synthesize them again. The determination of an empirical formula representing the proportions of the constitutive elements is certainly a necessary step in the synthesis of an organic compound, but it is a long way from being sufficient. For any complex organic compound, the synthesis is a veritable work of creation by the chemist.

From Simple to Complex

How can a chemist determine the appropriate pathways for an organic synthesis? For Berthelot, the art of synthesis was an extremely logical operation. He encouraged chemists to start from the basic building blocks of organic matter and then to move gradually step by step from the simple to the complex. On the Cartesian model, the problem had to be correctly decomposed, leading to Condillac's position — enthusiastically championed by Lavoisier — that to avoid error one needs always to proceed from the simple to the complex. Based on this principle, Berthelot introduced a grandiose programme that, he believed, would lead the chemist step by step from the four constitutive elements — carbon, hydrogen, oxygen and nitrogen — to all living matter.[8] For Berthelot, it was sufficient to proceed methodically, building up the desired compound one step at a time. Thus, the "art" of the chemist would not depend on inspiration or intuition, but would require the application of general rules and established methods. Berthelot envisioned this analysis as following four different stages:

1. Starting with carbon and hydrogen, one forms a series of binary compounds (the hydrocarbons) that constitute the backbone of all organic assemblies.
2. The tertiary compounds (alcohols) are formed by means of the appropriate affinities.
3. The alcohols can then be combined with organic acids to give ethers or the "aromatic essences", while the alcohols combined with ammonia give amines and vegetable alkalis. Aldehydes and organic acids are in turn formed by combining alcohols with oxygen.
4. Finally, the combination of organic acids with ammonium gives the amides (including urea).

The synthesis of certain well known naturally occurring organic compounds such as urea, gave Berthelot reason to believe that the way was already open for the synthesis of more complex compounds. Thus, he felt justified in proclaiming that synthesis "has eliminated the barrier between mineral and organic chemistry." He went on to argue that it had also demystified the vital force, implying that it was only the idea of this vital force that had justified the distinction between mineral and organic chemistry. In the end, his conclusion was simply that "the chemical effects of life are exclusively due to chemical forces."[9]

Berthelot went on to present a second argument that gave flesh to the idea that synthesis could be an effective tool for enhancing our cognitive powers. As he expressed it, "the experimental sciences have the power to realize their conjectures". Continuing in this vein, he compared the chemist's approach to synthesis with that of the mathematician.

> In their research into the unknown, both these orders of knowledge proceed by means of deduction. While the mathematician's reasoning, based on abstract data and established by definition, leads to conclusions that are both abstract and rigorous, the experimenter's reasoning, based on real data, which are always imperfectly known, leads to factual conclusions that

are never certain, but only probable, and so can never dispense with effective verification.[10]

Despite Berthelot's affirmation that chemical synthesis leads to less certain knowledge than mathematics, the chemist's capacity to form and test hypotheses is of the same order as mathematical deduction. The logic proposed for chemistry is not the empirical hit-and-miss strategy associated with inductive reasoning, but rather the hypothetico–deductive method. Thus, if the chemist proposes a general law, such as the existence of a series of intermediary bodies between two known series, then synthesis provides the means for verifying the conjecture. Berthelot cites the French organic chemist Charles Adolphe Wurtz as providing a good example of this kind of reasoning. Wurtz hypothesized the existence of an intermediate diatomic type of alcohol occupying the chemical space between the ordinary monoatomic alcohols, like ethanol, and the triatomic ones, like glycerine. In practical terms, he supposed that these intermediate alcohols could be formed from diatomic acids, and proposed to call them "glycols".[11] In this case, the technical manipulation that produced the material evidence served as the validating action, effectively producing the truth. Strikingly enough, this event was almost contemporaneous with the successful prediction of the existence of a hitherto unobserved planet by James Couch Adams and the French astronomer, Urbain Le Verrier. The hypothesized planet would, according to Le Verrier's calculations, explain the unexpected features of Uranus's orbit around the sun. This theoretical prediction based on calculations using Newtonian mechanics was confirmed in 1846 by the observation of Neptune in its predicted position by the Prussian astronomer Johann Gottfried Galle.

Thanks to the probative power of organic synthesis, chemistry could pretend to the same status as celestial mechanics. Indeed, thanks to experimentation, Berthelot saw chemistry as being capable of attaining the same degree of certitude as mechanics, with synthesis thereby serving to transform it into a "rational", predictive science. Once it was capable of making deductions from general laws, chemistry would finally transcend the chaos of empirical descriptions, where the properties of each individual substance had to be painstakingly determined and catalogued. Thus, for Berthelot, each synthesis was not simply the realization of a possibility, but the materialization of an idea.

While Berthelot's conception of a new rational creative synthesis may be seductive, it was too closely modelled on elementary chemical analysis to be anything more than a rationalist's dream. Despite dedicating decades to the task, Berthelot was not able to accomplish more than the first step in his universal project for organic synthesis.[12] More significantly, he never succeeded in developing any industrial syntheses, at a time when German chemists were laying the foundations of a fine chemical industry that would succeed in mobilizing organic synthesis to produce first tens and later hundreds of new compounds. Berthelot's synthetic chemistry was the mirror image of Lavoisier's analytical chemistry; from complex to simple, from simple to complex. While this approach appears quite rational, coherent, and offers a vision of total mastery over the processes involved, conceiving it as the inverse of elementary analysis would prove radically insufficient as a basis for practical organic synthesis.

From Fictions to Artefacts

For Berthelot, the chemical formula of a compound does not serve as a description of a hypothetical molecular reality, it is the result of an action, and offers a précis of the synthetic process, thereby constituting a "generative equation". Thus, Berthelot concluded that benzene was an isomer of acetylene because, when heated to 600°C acetylene produced a liquid containing traces of benzene, which he isolated in turn by fractional distillation. Nevertheless, this strategy had little future. By contrast, the chemists who based their research on August von Kekulé's hypothesis concerning the structure of the benzene molecule were able to produce a cornucopia of new and useful molecules. Chemical legend has it that in a reverie, Kekulé saw the six carbon atoms of the benzene molecule lined up in the form of a snake, and the snake then started turning in a circle until the head swallowed the tail on the model of the alchemical symbol of the Ouroborus. The six atoms in a circle, with alternate double bonds between them would form the basis of a wealth of artificial chemical compounds used in dyeing, medicine and the plastics industry. Indeed, the productive synthetic effort was based on substitution reactions, replacing atoms or functional groups in a given molecule, rather than the gradual composition of compounds starting from the original elements.

This approach to organic synthesis was opened up in the 1830s by another French chemist, Auguste Laurent, who rejected the dominant model of binary compositions. Inspired by contemporary research in electricity, Berzelius hypothesized a series of positively and negatively charged radicals, and supposed that any compound consisted of a combination of a positive with a negative one. Laurent believed that this approach had led to a science based on fantasy that had only succeeded in inventing a host of imaginary radicals. Indeed, he concluded in a famous polemic that "Chemistry today has become the science of bodies that do not exist".

When Bachelard cited this phrase from Laurent in his *Rational Materialism*, he likened it to Berthelot's formula describing chemistry as the science that goes from fictions to artefacts. Thus, Bachelard interpreted Laurent's approach as an extension of the ideal of a predictive and deductive science proposed by Berthelot. The substituted compounds are fictions, imagined before they are ever realized. Kekulé's benzene ring with its oscillating double bonds inspired the idea that there might be different isomers of di-substituted benzene.[13] Thus, this kind of reflection eventually led to Wilhelm Korner's successful isolation of the ortho, meta and para isomers of disubstituted benzene, which corresponded to the different orientations of the substitution groups around the benzene ring. Indeed, there are many ways in which theoretical prediction and practical experiment can interact in chemistry. For instance, Van't Hoff used the absence of isomeric forms as a proof for his hypothesis of the tetrahedral form of methane. If, as some chemists supposed, carbon compounds were planar, then one would expect cis and trans isomers of disubstituted methane compounds, depending on whether the two substitution groups were next to each other or on opposite sides of the carbon atom. Van't Hoff argued that no such isomers existed due to the tetrahedral form in which the substitution of two hydrogen atoms by the same group gives a single isomeric form. Thus, imagined molecules or "fictions" can serve to confirm a theory either by their presence or their non-existence. Peter Ramberg has argued that these fictional entities function in chemistry in the same way as "thought experiments" in physics.[14]

Whatever the role of these theoretical chemical fictions, the relationship between theory and experiment is hardly comparable to the linear deductions typical of mathematics. It is clear, for example, that analogy plays an important role in chemists' efforts to "realize their conjectures". It is rare for chemists to proceed in the linear style proposed by Berthelot. Instead they will search here and there for pieces of evidence to support a conjecture, often relying on plausible analogies to get ahead. This explains the importance of the literature organized around chemical analogues, such as *Chemical Abstracts* or *Beilstein*, as the reactions of and pathways to analogous compounds serve as the best guide for the chemist. This literature references literally millions of compounds, and is, as Peter Ramberg has suggested, comparable to reference works in legal systems where the jurisprudence is governed by the principle of precedence. Indeed, as in law, chemistry works by referring particular cases to one another, rather than the resolution of particular cases by the application of general rules.

Despite the argument we have just presented, once a claim has been confirmed by a successful synthesis, the singularity becomes a generality. As Bachelard has pointed out, the substances created by chemists are artefacts, even when they reproduce naturally occurring materials. Such substances come out of the organic chemistry laboratory in a state of purity never found in nature.

> One needs to bring into existence bodies that do not exist. As for the ones
> that do exist, the chemist must, in a sense, remake them in order to endow
> them with the status of acceptable purity. This puts them on the same level
> of "artifice" as the other bodies created by man.[15]

The creation of artefacts — in the literal sense of objects made or manufactured by human beings — is not, therefore, simply a technological by-product of chemical research. Before becoming industrial products, these artefacts served as a means for establishing chemical knowledge. Even today, synthesis is not exclusively a way of satisfying technological demands, but remains a privileged tool for understanding the nature of chemical compounds. Indeed, the majority of our knowledge concerning the molecular structure of organic compounds derives from nineteenth-century syntheses, starting with molecules like alizarine and indigo and including Emil Fischer's synthesis of polypeptides and tannins. The techniques developed to create these products were not, however, limited to these initial applications. The various substitution reactions provided the means for generating a potentially infinite number of compounds, of which, inevitably, only a minority had any naturally occurring analogues.

Organic synthesis necessarily implied a dialogue between theory and practice, reflecting its potential not only for exploring the configuration of particular molecules but also for verifying quite general theories. Woodward considered that this exchange between theory and practice constituted the basis of a "second revolution" in organic chemistry. This followed a "first revolution" introduced by the structural theories proposed by Laurent and Kekulé.

> It is our view that organic chemistry has just passed through its second great revolution. The structure theory recognized that the maintenance of nearest-neighbor relationships among elements was responsible for the variety and individuality of the material components of the physical world. The great advance of the recent past has been the recognition of the entities responsible for the maintenance of those near-neighbor relationships, and a description in simple general terms of wide applicability and precision of their fluid nature, and of the laws to which they are subject. The resulting edifice of organic chemical theory enables us, with obvious consequences for organic synthesis, to assert that the outcome of very few organic reactions is unexpected, and fewer inexplicable.[16]

Beyond the considerations of molecular structure of organic compounds, synthesis has played a vital role in much more fundamental theoretical developments. Indeed, the determination of energy values associated with molecular and atomic orbitals that can be estimated using Schrödinger's equation can only be verified by means of carefully controlled syntheses which are capable of generating measurable indices. Nevertheless, all these organic syntheses are founded on a representation of molecular structure, and are not at all conceived of as being simply analysis in reverse. These syntheses mobilize a projected world constructed by human thought that remains fictional until the synthesis has been successfully completed.

Of course, with the rise of digital molecular modelling, these fictions often assume a virtual form as well. After conception, the organic chemist operates with the available tools to construct the molecule in question, adding and removing groups as if operating with an elaborate "Meccano" kit. The ability to use this kit composed of standard reagents along with useful reactions or specific physical conditions conducive to one orientation or another cannot be reduced to an overarching chemical theory or even a well-defined set of general laws.

A Creative Process

Theory plays more than one role in the domain of organic synthesis. In addition to helping the chemist to plan the series of reactions that will lead to the desired compound, it also serves to indicate potential obstacles or pitfalls. How can we synthesize compounds that are theoretically possible? Indeed, as we have just been arguing, the creation of an "artificial" molecule is not a simple process of deduction from a theory or even a set of theories. Chemistry does not comply with the deductive methodological model that stands at the centre of traditional philosophy of science. While the synthesis of hundreds of benzene derivatives at the turn of the nineteenth and twentieth centuries depended on the hypothesis of the hexagonal ring structure of the benzene molecule, it also required an experimental innovation, the introduction of a suitable catalyst by Friedel and Crafts in 1877. The use of anhydrous aluminium chloride allowed the chemists to substitute an alkane group for hydrogen on the benzene ring using the appropriate organic chloride. These kinds of reactions, which provided the steps for organic synthesis starting at the end of the nineteenth century, still bear the name of their inventor — like the Grignard reaction — and constitute veritable inventions on an industrial model. With a large number of applications, particularly in the artificial dye industry and later in pharmaceutical research, the economic stakes of these inventions were of the highest order, and they were often protected by patents.

We do not want to claim that structural theory has not had a decisive influence on synthetic organic chemistry, but it did not by itself provide the solutions to the practical problem of how to synthesize any given compound.[17] It was the artificial dyes, in particular mauve — synthesized by

William Perkin — that marked the entry of synthetic organic chemistry onto the industrial scene. Perkin discovered mauve as part of his abortive attempts to synthesize quinine in 1856, shortly before the widespread introduction of structural formulae. Thus, this landmark invention was clearly not planned in advance, either in terms of its chemical structure, or its properties. How many other useful materials — the glue for Post-Its, Nylon, or Gore-Tex to name but three — were the accidental products of chemical tinkering? As the research for artificial dyestuffs intensified, particularly in Germany at the end of the nineteenth century, many teams of chemists were mobilized in the first industrial research laboratories, with companies investing large sums of money in facilities, equipment, staff, and the management of intellectual property rights. This development has characterized the evolution of the chemical industry and its offshoots (pharmaceuticals, plastics, etc.) right up to the present day.[18]

Organic synthesis, particularly in its industrial form, is far from being a simple "application" of advances in theoretical science. Indeed, among the wide range of resources mobilized in this domain, relatively few are theoretical in the traditional sense of the term. To attempt a preliminary and approximate taxonomy of these resources, we suggest dividing them into three categories, depending on whether they are implicated in analysis, synthesis, or scaling up.

First, all industrial production involves some kind of analysis of the end product in the interests of quality control. These analytical techniques serve to verify the composition of the intermediary substances at each step, as well as standardizing the end product. Analysis of a competitor's product can also serve in the quest to imitate or improve upon it. While quality control has become increasingly dependent on technological devices, this was not always the case, with a skilled foreman in a steel foundry, for example, having the ability to tell at a glance if the right quantity of carbon had been added to the molten iron. This judgement depended on the appearance of the liquid and in particular its colour. Such informal analytical techniques were the result of empirical experience rather than the molecular theories of materials science.

Moving on to our second category, it is clear that synthesis requires the mastery of a whole range of procedures taught to industrial engineers. While this training is largely theoretical, the choices of which particular procedures to use in practice, or the invention of new ones required for new production processes always implies a certain amount of empirical experimentation based on a sophisticated process of trial and error.

Finally, scaling up concerns the techniques required to transfer a series of reactions from the laboratory bench to a feasible industrial process for large-scale manufacture. The principle is a straightforward one, but the scaling up process has often posed more problems than the development of the synthetic pathway, and many inventive syntheses have turned out to be unfeasible on an industrial scale. On the other hand, ingenious procedures for large-scale production have revolutionized the chemical industry, such as the Leblanc process for artificial soda or the Haber process for the production of ammonia. Scaling up is an interdisciplinary process, drawing on skills in physics, mechanics, economics, and architecture, because devising a competitive production process requires all these skills and more besides. A training as an organic chemist, let alone as a theoretical organic chemist, is insufficient for the chemical engineer responsible for this process. Writing in 1807, Chaptal, chemist, entrepreneur, and later a successful politician, had already recognized that the introduction of an innovation into the workshop required a great deal of caution.

> The manufacturer could all too easily compromise his reputation if he were to determine his conduct by, or found his speculations on a few laboratory results or deceptive appearances. It is only with the greatest circumspection that one can import innovations into the workshop, however profitable they may appear to be.[19]

The dominant image of technology and industry guided by theoretical science that is promoted by industry and scientists alike hides a less rational, more craft-like reality. Much of the knowledge needed for the synthetic chemical industry is not learned at school or university. This is not the "universal" knowledge of scientific theories, but rather "local" knowledge, or knowledge of "localities" as Chaptal termed it. For Chaptal, crucial factors included the geographic and human environment of the factory, but also extended to tacit knowledge learned on the job, much of it by trial and error, with accidents often serving as object lessons. Indeed, since the time of Chaptal, chemical engineering has emerged as an attempt to rationalize chemical production with the application of physics and mathematics to the process of transforming raw materials into chemical products. By the end of the nineteenth century, courses in chemical engineering started to be taught in a number of universities and process engineering has since became a specialty in itself.

In light of this interpenetration of laboratory work and industrial production, it should be easy to see that the schema of pure and applied chemistry is illusory in a number of respects. In particular, it presents an image of a one-way flow from the development of a synthesis based on the theory of the pure science to its application in an industrial production process. Even before this distinction was declared obsolete by the rise of the biotechnologies and other approaches that supposedly blurred the boundaries, a two-way exchange between industrial practice and university research had always existed. A large number of nineteenth-century chemists received their training as apprentices in one of the chemical arts, and while generally not a source of pride for university chemists at that time, many supplemented their incomes as consultants for industry.

While philosophers of science have often been seduced by the parallel between the two oppositions, theory versus practice on the one hand and pure versus applied science on the other, academic chemists, even at the highest levels of the science, have been more sceptical about the distinction. Thus, in the 1990s, the Nobel-Prize-winning chemist Roald Hoffmann presented a series of reflections entitled "in praise of synthesis" that are in some sense similar to Berthelot's thoughts from a century earlier. "Synthesis is a remarkable activity that is at the heart of chemistry, that puts chemistry close to the arts, and yet has so much logic in it that people have tried to teach computers to design the strategy for making molecules."[20]

Nevertheless, the similarity ends here, as Hoffmann goes on to evoke a multitude of different types of synthesis. He points to a profound difference between planned industrial synthesis dominated by economic considerations, and fundamental laboratory research governed by chance, where researchers constantly mobilize their intuitions and the tricks of the trade that will allow them to overcome what others might perceive as insurmountable obstacles. The ingenious chemist is capable of using the laws of nature to achieve results that appear to be excluded by these very laws. Hoffmann likens synthetic chemistry to a game of chess played against nature itself. The chemist moves through the reaction pathway step by step; breaking a bond here to form another one there, catalysing one reaction to generate a necessary intermediate compound while ensuring that its yield is favoured by the reaction conditions, directing substitute groups to a specific site while blocking unwanted transfers or reactions that could take the intermediate product off down the wrong pathway. This kind of wily game with nature is the art of the synthetic chemist. Each new molecule is a personal or collective victory demonstrating the chemist's skill and imagination. The world in which chemists generate new molecules is not a world of theory, but a world of material interactions, governed as much by intuition and imaginative innovation as by the systematic use of theoretical knowledge and logic.

References

1. M. Berthelot (1876), p. 275.
2. R. B. Woodward (1956) and J.-M. Lehn (1995), in the preface.
3. M. Berthelot (1876), p. 275–276.
4. *Ibid.*, p. 277.
5. *Ibid.*, p. 269.
6. *Ibid.*, p. 35.
7. *Ibid.*, p. 66.
8. While Lavoisier regarded organic compounds as composed of carbon, hydrogen and oxygen, many of his collaborators thought nitrogen could also enter into such compounds. Despite claims by chemists such as Vauquelin to have identified various other elements like phosphorous, arsenic or iron in animal and vegetable matter, these claims were far from being accepted by all chemists.
9. M. Berthelot (1876), p. 272.
10. *Ibid.*, p. 276.
11. *Ibid.*, p. 190–192.
12. Berthelot's most notable achievements included the synthesis of formic acid in 1856, methane in 1858, and acetylene in 1862. He had previously synthesized benzene in 1851 by heating acetylene in a glass tube.
13. The three isomers are those with the two groups next to each other — ortho or 1,2 — with the groups separated by one carbon atom — meta or 1,3 — or separated by two carbon atoms — para or 1,4.

14. P. Ramberg (2001).

15. G. Bachelard (1953), p. 22.

16. R. B. Woodward (1956), p. 156.

17. See, A. S. Travis, *et al.* eds (1992).

18. Thus, the pharmaceutical industry was one of the first to invest heavily in research and development and continued to do so over the course of the twentieth century. This model of research-intensive industry established in the chemical industries has since spread to many other sectors, such as computing, aeronautics, etc.

19. J.-A. Chaptal (1807), p. xv.

20. R. Hoffmann (1995), 'In praise of synthesis,' p. 95.

CHAPTER 7

A DUEL BETWEEN TWO
CONCEPTIONS OF MATTER

The Ancient Concept of *Phusis*

Like the related doctrine of the elements, the theory of atoms was first elaborated in ancient Greece long before chemistry took its modern form. While there was no doubt considerable practical knowledge of a range of chemical reactions (particularly, useful transformations of matter), this was mainly a question of skill-based, tacit knowledge, confined to the practitioners of what came to be known as the chemical arts, such as dyeing or metallurgy. The first theories of matter were not elaborated in the context of these chemical practices, but in response to the abstract issue associated with the concept of "*phusis*". Although this word lies at the origin of the contemporary term physics, *phusis* embraced a large ensemble of problems concerning the essential nature and functioning of the material world. It was in this philosophical context of *phusis* that both the theory of elements and the theory of atoms were developed. Thus, they not only presupposed but also helped to construct a shared conception of nature that has marked the history of philosophy up until the present day. These two philosophies share two fundamental premises: first, the principle of the conservation of matter; and second, the idea that the world should be conceived of as being primarily phenomenal, in the sense of being a collection of phenomena.

Most students of chemistry have diligently learned that it was Lavoisier who introduced the concept of the conservation of matter into chemistry, using this principle to dismiss the imaginary element of phlogiston. Armed with this concept of the conservation of matter, he could argue that the weight gained by a metal in the process of calcination (today known as oxidation thanks to the work of Lavoisier and his collaborators) could not be explained by the loss of phlogiston, as this would mean that phlogiston would have to possess a negative weight. The consistent solution was that a material was absorbed in the process. This mysterious matter was gaseous oxygen, which had escaped the attention of scientists prior to the elaboration of pneumatic chemistry. Nevertheless, the conservation of matter is a basic assumption that underlies ancient physics. A great majority of Greek philosophers and early modern scientists considered matter to be eternal and indestructible without having any experimental evidence for it. The conservation of matter is so deeply embedded in western science that the philosopher Emile Meyerson considered it as an *a priori* metaphysical assumption and the necessary foundation for all scientific endeavours.[1] In reality, the question that haunted ancient physics was not the origin of matter, but rather how an original chaos was transformed into the ordered universe that we inhabit, or the "cosmos" to use the Greek term. Over time, this issue ceded its pre-eminence to another one: how there can be permanence or at least continuity in a world of constant flux, or, in other terms, what is the nature of identity in the context of change?

The conception of the world as a collection of phenomena derives from our nature as sensorial beings. It is easy to recognize, thanks to the tricks of the light or echoes, that our perceptions can transmit a false impression of a world that we can sometimes know by other more reliable means to be different from the way we perceive it. Thus, there was a sense that the quality of human knowledge of the world was inferior when compared to that of an omniscient being or beings, who were capable of knowing the world as it is. The theories of atoms or those of elements aimed to offer a description of the permanent underlying reality. Nevertheless, even for the atomists, humans were incapable of directly experiencing this real but invisible world of atoms, and would have to be content with deriving their knowledge about it from observing perceptible phenomena. It was the Enlightenment philosopher, Immanuel Kant, who devised the vocabulary in which the noumenal world, derived from the Greek word *noumenon* (the thing in itself), is contrasted with the phenomenal world (i.e., the world of sense experience, a term derived from the Greek verb *phainein* which means to show).

The latin term "element" serves only as an approximate translation for a relatively complex concept forged in ancient Greece.[2] Element has served to translate the word *arche*, used to denote the primitive principles in the first cosmogenies, such as water or the principle of moisture, which features so prominently in the writings of Thales of Miletus. Thales conceived the cosmos as originating from a sort of primitive ocean, with the variety of substances present on the earth today separating out by the processes of rarefaction and condensation. But element has also been used to translate Empedocles' *rhizomata*, in the sense of the roots of material things, and to translate Aristotle's four "elements" that he denoted by the term *stoicheia*. Finally, element was sometimes used by the Roman poet Lucretius in his epic defence of Epicurus's atomism *De Rerum Natura*. Here, the element designates the indivisible or inseparable units otherwise known as atoms (from the Greek *a-tomos* meaning indivisible). Thus, it would be naive to characterize the opposition between the philosophy of elements and that of atoms simply in terms of the two concepts. Instead, we should regard them as two rival systems proposing two different styles of explanation for the phenomena of nature.

Atoms versus Elements; Two Rival Systems

Without entering into the details of these two systems, we can offer a quick description of the theory of elements that differentiates it from atomic theory. In the elemental approach, the fundamental "principles" are considered to be entities characterized by their specific inalterable properties. Thus the four "roots" described by Empedocles of Agrigentus — earth, air, fire and water — come together under the dominion of love and separate under the rule of hate. The principal function of these incorruptible, eternal elements is to guarantee permanence in a world of perpetual flux, and unity in a context of unrestricted variety. The material world results from a mixture of these elements in diverse proportions, just as a painter is capable of concocting an endless variety of colours from the four primary colours she finds on her palette.[3] Empedocles seems to suggest, however, that prior to the existence of the four elements, matter was composed of infinitely small, identical entities — the minima. Commentators on Empedocles, including Aristotle, referred to these "minima" as "homeomers", drawing particular attention to the idea that all their parts were alike, and treated them as a sort of "raw" matter out of which the elements (themselves clearly differentiated) were constituted. For Aristotle, each of the four elements was formed by the union of a formless material substrate

FIGVRA XXXIV.

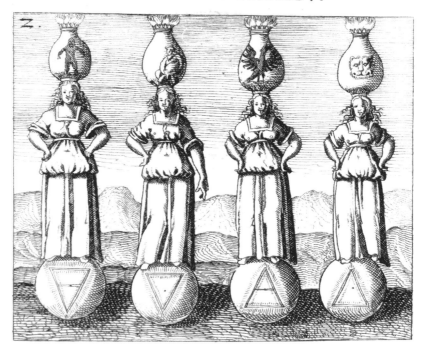

Figure 8. The four operations of chemistry — solution, ablution, conjunction and fixation — represented as four sisters. The female figures also make reference to the four Aristotelian elements: earth, water, air and fire (left to right). Daniel Stolcius von Stolcenberg, *Viridarium chymicum*, Frankfurt, 1624. Reproduced with the permission of the Bibliothèque nationale universitaire de Strasbourg.

(the substance) with two of the four essential qualities. Thus, each element materializes two qualities: earth is cold and dry; water is cold and humid; air is hot and humid; and fire is hot and dry; although one quality predominates in each element. The association of qualities with a substrate devoid of any such properties allows the Aristotelian elements to function as vehicles for these qualities. Fire bears lightness, as it is in its nature to move away from the centre of the world, while Earth bears heaviness, as it is in its nature to move towards the centre of the world.[4] In other words, the quality is inherent in the element, and so when an element enters into a mixt, it confers its dominant quality on the mixt. Indeed, in this picture, we cannot really talk of elements "constituting" Aristotelian mixts in the same way as we would today say that hydrogen and oxygen constitute water. In Aristotle's view, the four elements possessed at least one quality to an extreme degree — dryness for earth, heat for fire, etc. — while the mixts possessed the same qualities, but necessarily to a lesser degree.[5]

This interpretation of the ultimate nature of matter was periodically attacked, particularly by the champions of the atomic theory developed by the school of Abderus and reformulated by Epicurus and his best known popularizer, Lucretius. Followers of this school refused to accept the type of explanation that was put forward by Empedocles. In his first book, *De rerum natura* (*On the Nature of things*) Lucretius laid out the four principal reasons for not accepting Aristotle's approach: (1) it supposes there can be movement without accepting the existence of any void; (2) these philosophers place no limit on the division of matter, meaning that the material world must ultimately disappear; (3) air and earth, and water and fire are opposites and mutual enemies, implying that each one should annihilate its opposite, and so they cannot guarantee the permanence of matter; (4) last, but most important, the cycle of transformations that generated air out of earth and water out of air demonstrates that all types of bodies can generate all the others, meaning that none can truly qualify as a principle.[6] Thus, it was inconceivable to the atomists that the world could be composed of mixtures of a small number of "essential" constituents in varying proportions. Empedocles had presented this position in terms of an analogy with the painter's palette, with four primary colours giving rise to an infinite range of colours, exhibiting all degrees of subtlety. By contrast, Lucretius argued that such a system applied to the material world could only give rise to disparate heaps of matter, involving the simple apposition of the essential principles rather than well-defined, differentiated bodies.

The atomist vision of the world posited its being constituted by a single homogenous matter distributed in tiny, invisible, solid, eternal, indivisible units continuously in motion within an immense void. The infinite variety of bodies that make up the material world is due to the concatenation of these atoms in various arrangements, just as one can compose an infinite variety of texts by arranging the limited number of letters of the alphabet in different combinations.[7] The metaphor of the letters of the alphabet suggests three principal differences between the atomists' view and an elemental one. First, the physical world is composed of discrete units of matter that differ only in terms of their form or shape and not their properties or nature as is the case with elements. This ontological contrast results in different forms of explanation. Second, according to the atomist, the only factors determining the generation and the properties of sensible bodies are the shapes and positions of the atoms. Consequently, they provide mechanical explanations in terms of diverse connections, weights, shocks, encounters and motions of invisible atoms.[8] Finally, the alphabet metaphor suggests that there are rules, codes, or laws that govern the combination of atoms. Such laws, moreover, are ultimately responsible for the nature of the sensible world.

While to our eyes this doctrine of atomism, at least in the very general terms in which we have presented it, might appear more modern than the doctrine of the elements, it nevertheless represents just one possible way of interpreting nature, and it would be a great historical mistake to think of it as essentially "correct." Rather than re-describing the history in terms of who was right and who was wrong, it is more fruitful to consider the confrontation of these two doctrines as just that: a confrontation between two doctrines. One doctrine postulates the existence of minimal units, discrete particles of homogeneous matter, and considers that all sensible qualities can ultimately be reduced to the movements of these atoms and the forms assumed by their concatenations. The doctrine of elements considers matter to be continuous although heterogeneous and well differentiated. The first of these doctrines is oriented towards a geometric physics that invokes purely mechanical explanations, while the second suggests the elaboration of a qualitative physics, a science of "mixts" rather than aggregates of individual bodies. Having distinguished these two doctrines developed in the Ancient Classical World — atomism and elemental theory — we can now ask whether these two traditions are the sources of the two modern disciplines of physics and chemistry, respectively.

Is Chemistry Aristotelian?

Two characteristics of the Aristotelian four-element theory of principles seem to support an argument for its continuity with modern chemistry. First, there is its characteristic pluralism, which suggests that while we can reduce the diversity of natural phenomena to a few irreducibly differentiated entities, there is not just one single universal type of matter. Nevertheless, the phenomena we observe at the macro level still find their explanation in a mixture of the ultimate elements. As Auguste Comte pointed out, the number of elements we are prepared to accept is going to determine their nature. Thus, the shift from one ultimate principle to four represents such a profound change of perspective that Comte was ready to interpret it as "the veritable origin of the chemical sciences". Moving from a single element to multiple elements, the scientist is forced to abandon the notion of the absolute, an essential precondition for conceiving of composition and decomposition.

Comte viewed this "revolution", which he attributed entirely to Aristotle, as much more significant than the subsequent evolution that led "by gradually improving exploration, from Aristotle's four elements to the fifty six simple bodies of today's chemistry."[9] Thus, for Comte, chemistry was unquestionably a modern descendant of the Aristotelian tradition, with Boyle and Lavoisier representing just a pair of milestones on what was essentially a continuous path of discovery elaborated within the same conceptual framework. Whatever its historical weaknesses, this position at least has the merit of highlighting the fundamental difference between two ancient philosophical senses of the term "element". Aristotle did not adopt the sense of element as the origin of all things, but as something that circulates in the course of metamorphoses and other transformations, while itself remaining essentially unchanged.

The second aspect of Aristotelian elements that potentially supports the thesis of continuity is the idea of qualities. For Aristotle, qualities were the elements' enduring attributes that served to define their ultimate nature. This concept of qualities has served two different functions for the chemist, one classificatory and the other explanatory. On the taxonomic side, the fundamental properties associated with qualities supply a set of generic principles that can be used to group individual substances in classes. The idea of inalienable properties that depend on an element's qualities serves to explain many common chemical phenomena. It is well known, for example, that substances with certain properties can be combined to obtain new substances with different properties, and yet the initial substances can also be subsequently recovered in their original form possessing the same qualities as before.[10]

These very general "filiations" between Aristotle's matter theory and modern chemistry, or even medieval alchemy offer, however, no more than a vague genealogy. Moreover, they can be misleading, because the physics of the ancient Greeks was not elaborated in response to any particularly chemical questions, making it unsurprising that it was largely inadequate as a framework for conceptualizing the chemical sciences. Already by the Renaissance, there was an important body of experimental (and theoretical) chemical knowledge in Europe that fit poorly with Aristotelian conceptions. Thus, for a long time, even though ever more chemical transformations were known and empirically mastered, they remained loosely theorized in a rather eclectic manner. Despite these problems, we can still ask whether it makes sense to argue that medieval alchemy followed this Aristotelian philosophical approach, at least to a certain extent.

The Aporia of the Mixt

For Pierre Duhem, chemistry was the heir to Aristotelian philosophy because Aristotle had not only stated the central problem for this science but had also proposed a solution for what he characterized as the enigma of the mixt. In his *On Generation and Corruption*, Aristotle considered the issue of the genuine mixing of different bodies, a possibility entirely denied by "certain philosophers".[11] Indeed, if the components of a mixture remain intact, like the grains of barley and wheat in a mixture of different cereals, there is no genuine mixing in the chemical sense. While such a mixture might represent a "mixt to the eyes" for a normal person, the heterogeneity of the mixture could not escape the "eye of the Lynx". If, however, the original bodies combined together no longer exist but have been replaced by a different body — the mixt — one has to contend with the fact that a new body has been generated. Thus, the phenomenon of a "mixture" is either a case of aggregation — involving the juxtaposition of discrete units of the different components — or a case of generation and corruption. The concept of the "mixt" plunges us into a paradox; either it is the original constituents sitting side by side, or else something genuinely new has been created that does not possess the properties of the original ingredients. But the emergence of a new material implies that the ingredients no longer coexist within the mixt. Consequently, a true mixt can be characterized by an either...or condition. Either the mixt is in the state of a "compound", and the properties of the initial ingredients are lost, or the original ingredients are recovered and the properties of the mixt are lost.

Aristotle avoided this apparent aporia by distinguishing between the potential and the actual in such a situation. While the elements that enter into the mixture may cease to exist in their actual states as elements, they retain the potential associated with their nature as elements, meaning that they continue to exist (in their potentiality) in the genuine mixture, even though their presence is (for the duration of the mixture) undetectable.

It was Aristotle who coined the term *synthesis*, with the literal sense of placing things together, or juxtaposing them. Albertus Magnus later translated the Aristotelian term "*synthesis*" into the latin word "*compositio*". Originally, therefore, *compositio* literally signified the same thing, that is to say an apparent mixt and not a genuine one, a "mixt to the eyes" in the sense explained above. Likewise, in its original sense, synthesis does not imply the production of a new substance different from the reactants by a chemical interaction. Nevertheless, over the course of the centuries, this term (specifically in the form of "compound") came to replace the term "mixt", which was the translation of Aristotle's *mixis*. The fact that "mixt" ceded its place to "compound" suggests that modern chemistry is not really the heir to Aristotelian philosophy.

Mixts or Compounds, Stahl or Lavoisier

At the beginning of the eighteenth century, Georg-Ernst Stahl echoed Aristotle's distinction between true and apparent mixts in an attempt to delineate the territory of chemistry from that of physics. While mechanical physics could account for "aggregates", only chemistry could handle "mixts". Aggregation was a juxtaposition of units, and could be understood in mechanical terms such as mass and movement, but mixtion was the union of principles involving individual affinities. The decomposition of an aggregate would not affect its properties whereas mixts could only be analysed by changing their properties. With this approach, Stahl centered chemistry on the notion of the mixt, defining chemistry as "the art of resolving mixt, compound or aggregate bodies into their principles; and of composing such bodies from those principles."[12] Stahl assumed that mixts were made of four qualitatively different primary bodies consisting of three different kinds of "earth" and water. It seems that unlike Aristotle, Stahl considered that the mixts actually contained their constituent primary bodies. So successful was this conceptual strategy that many in the eighteenth century considered Stahl the founder of chemistry.

In the eighteenth century, however, chemists stopped using the word "mixt", as it was replaced by the notion of composition or compound. In particular, Lavoisier's famous definition of elements as uncompounded substances was an integral part of a reorganization of chemistry along the lines of another distinction, the one between simple and compound. Lavoisier, who came to be celebrated as the founder of "modern chemistry", redefined chemistry as the science concerned with decomposing the natural bodies and "examining separately the various substances entering into their combination". While the compositional perspective was not new, it became the dominant paradigm following the reform of chemical language.[13] In the new language, published in 1787 by four French chemists, Guyton de Morveau, Lavoisier, Fourcroy and Berthollet, names of compounds were coined by simple juxtaposition of the names of the components, and were considered to offer "mirror images" of the actual composition of material bodies.[14] Lavoisier, who admired and extensively quoted Etienne Bonnot de Condillac, in particular his *Logique* from 1780, adopted Condillac's view of languages as analytical methods as well as his notion of analysis as a two-way process, from simple to compound and from compound to simple. According to Condillac, analysis is a mental process that allows one to visualize a picture presented simultaneously to the senses, as a succession of elements. Condillac developed a metaphor of someone glimpsing a view from the window of a castle. This glimpse allows the observer to see the whole scene but remains imperfect as a source of knowledge. It is only the subsequent detailed observation or analysis that brings true understanding. With this analytical regard, the observer distinguishes and names every element of the landscape, ordering each one in relation to the others. Although all this information was present in the initial global view, it is only analysis that can provide a true understanding of the scene.[15] Condillac's logic, inspired by algebra, in turn inspired not only Lavoisier's collaborative language reforms, but also the use of equations to describe chemical reactions.[16] A compound is described as the addition of two constituent elements, and is entirely characterized by the nature and proportion of its constituents. The use of the sign "equals" in the equation clearly indicates that chemists no longer care about the either/or condition: the two sides are considered equal even though the reactants are quite clearly different from the products. The puzzling issue raised by Aristotle about the mode in which the constituent elements are present in a compound was put to one side by Lavoisier, who found that ignoring it was more practical than solving it.

The compositional paradigm proved very successful when reinforced by John Dalton's atomic hypothesis. By the middle of nineteenth century, the definition of a compound by the nature and proportion of its constituents was being challenged by a structural paradigm that took into account the arrangement of atoms in molecules. Nevertheless, structural formulas do not respect the either/or condition any more than the compositional approach, but instead displace it. On this model, it is the physical arrangement of the constituent elements that accounts for the properties of the compound.

A Vexing Question

While we are entitled to conclude that modern chemistry is not a direct descendant of Aristotle's philosophy, Pierre Duhem believed that it was, and it is worth taking the time to examine his reasoning on this question. Duhem's thoughts on chemistry were published in a book entitled "*The Mixt and Chemical Combination*" (*Le mixte et la combinaison chimique*), an old fashioned title, particularly when compared to the one he chose for his book on physics *The Aim and Structure of Physical Theory* (*La théorie physique, son objet, sa structure*).[17] Why, we might ask, did he choose to use the term "mixt" in 1902 when it no longer formed part of the contemporary chemist's vocabulary? A century earlier, the debate between Joseph-Louis Proust and Claude-Louis Berthollet concerning the nature of chemical combinations had given rise to a new consensual terminology. Henceforth, the (mechanical) combination of chemical species in indefinite proportions would be termed a "mixture", while their chemical combination in definite proportions would be called a "compound". Thus, Duhem deployed the word "mixt" as a self-conscious evocation of an outmoded chemical concept, in order to investigate its philosophical associations. Duhem introduces his topic by describing a familiar empirical observation that doubles as a simple chemical experiment.

> Throw a little sugar into a glass of water. After a short time, the solid, white crystalline body which constitutes the sugar has disappeared. The

glass contains no more than a homogeneous liquid, transparent like water, but with a different taste. What is this liquid? The vulgar call it sugared water. The chemist says that it is a solution of sugar in water. These two descriptions correspond to two essentially distinct opinions.[18]

Based on his analysis of this experiment, the alternative between atom and element discussed above is introduced by Duhem in the form of a difference in interpretation. The common-sense interpretation of the experiment is based on the sensorial appreciation of the phenomenon — the colour, taste, and odour of the sugar-water — while the chemist looks beyond this kind of immediate sense-experience to evoke the dissolution of the sugar molecules in the water. While for Bachelard, this example illustrates precisely the epistemological obstacles that the scientific spirit has to overcome in order to escape the limitations of such a common-sense interpretation, Duhem approaches the issue from the opposite direction. For Duhem, the common-sense view is the correct one, as the chemist has made an unjustified leap from a macroscopic to a microscopic interpretation of the phenomenon. The chemist's interpretation does not rest on an examination of the details of the experiment, but instead mobilizes a molecular imagery, which, according to Duhem, proceeds from the imagination rather than sensation. The common-sense denomination implies that the non-scientist has the courage to confront the enigma of the "mixt": "the glass no longer contains either the water or the sugar that we placed in it, but a new body, a mixt formed at the expense of these two elements". In order to resolve this enigma, Duhem explicitly deploys Aristotle's conceptual apparatus: "While the sugar–water no longer currently contains the water and sugar that served to form it, it can, by ceasing to exist, reproduce this water and this sugar: it contains them in potentiality."

This preliminary reflection concerning sugar-water was not intended to promote the value of common-sense opinion but to challenge the molecular vision of modern chemists. Duhem was inviting the contemporary chemist to question the adequacy of the dominant paradigm represented by molecular models. He revived the notion of the mixt because he believed that atomism and molecular architecture, the approaches that dominated organic chemistry at the beginning of the twentieth century, were incapable of providing an exhaustive explanation of chemical transformation. Thus, despite undeniable practical and theoretical progress in the science, the central philosophical problem confronting the chemist was the same as that faced by Aristotle two thousand years earlier. A mixt is the product of two or more components that disappear in the process to form a new entity, and yet it is possible to decompose this mixt to recover the original components. Rock salt possesses completely different properties from either chlorine or sodium, even though it is capable of generating these two elements that enter into its composition. Therein lies the enigma of chemical composition; the conservation of matter accompanied by the emergence of novelty. The properties of the mixt are never the simple sum of the properties of its components.

While Duhem did oppose atomism, he did so for quite different reasons than other famous French chemists like Henri Sainte-Claire Deville and Marcellin Berthelot. Based on a positivist principle that science should limit itself to observable facts, these French anti-atomists condemned the approach because it went beyond experience. Although Duhem discussed this argument, it was not his principle criticism, as he had no intention of limiting scientific theory to talk of observables. Indeed, again following Aristotle's lead, Duhem thought the chemist would have to think in terms of constituent elements that were invisible even to the Lynx-eyed observer or the most powerful microscope possible. Updating this argument to a contemporary setting, we could say that nothing observed by either the electron microscope or the scanning tunnelling microscope, or any other visualization technique, is capable of explaining the phenomenology of the macroscopic behaviour of physical bodies.

The problem that Duhem sees with atomism is, therefore, not so much its status as a metaphysical hypothesis, but its weak explanatory power. An examination of atomic notation provides Duhem with the means to clearly pose the problem of the relationship between the sensible properties of a compound and the nature of its constituent elements. This critical analysis targets not only atoms but also elements in the sense of simple bodies that explain the nature of the compound. The typical reasoning of a post–Lavoisian chemist is to explain the sensible properties of a compound by reference to the nature and the proportion of its constituent elements.[19] Duhem takes the opposite tack by suggesting that the compound should be used to explain the properties of the element. Valencies should not be considered as the invariable intrinsic properties of each atom (thus, Duhem refused to use the contemporary term "atomicity") but rather as properties of a particular compound. Using what he had learned from Aristotle's philosophy, Duhem rejects the choice between atom and element, thereby escaping the stifling to-and-fro between simple and composed. The mixt — the phenomenal compound body that the chemist has to deal with — cannot be reduced to elements or to atoms. It is this concept of irreducibility placed at the heart of chemical theory that justifies Duhem's use of the outmoded term "mixt" in the presentation of his philosophy of chemistry.

In what follows, we will have more to say about how Duhem translated and exploited the Aristotelian notion of potentiality or disposition within his own philosophy of science. For now, however, the essential point we should retain from this discussion is that as long as chemical theory, or chemical practice for that matter, presupposes the possibility of establishing a causal relationship between properties and composition, there will always be two different interpretations available to explain such a relationship — the elementary and the atomic in the sense described above. Thus, the belief that modern chemistry is the direct heir of the ancient theory of elements is either the result of a superficial analysis, such as the one offered by Comte, or the result of a partisan philosophical position, as is the case with Duhem. Rather than feeling obliged to choose one over the other, we prefer to see this duel between the two approaches as a characteristic feature of the history of chemistry. Chemists have always been confronted with this interpretative dichotomy, and, depending on the period, they have opted for a version of atomism or an elementary approach, or else have tried to reconcile the two. Nevertheless, chemistry continues to be haunted by this interpretative plurality, which orients us towards a profound philosophical question. It is not, however, our intention here to resolve the issue, merely to indicate the co-existence of the two approaches.

References

1. E. Meyerson (1921).
2. One proposed — but controversial — etymology for element suggests that it originated from the alphabetical sequence of letters l, m, n (like 'alphabet'

itself, which comes from the first two letters of the Greek alphabet, alpha and beta). One of the appeals of this etymology is that the Greek word 'stoicheion' was originally used to designated objects arranged in a series, like the letters of the alphabet.

3. Empedocles, Fragment 23.

4. Since fire and earth tend to move up and earth and water to move down, Aristotle assigns a specific space to each of the elements in the sublunar world. Thus the sub-lunar world is essentially composed of concentric spheres of earth, water, air and fire.

5. P. Needham (1996).

6. Lucretius (50 BCE.) book 1, 635–920.

7. *Ibid.* book 1, 820–29.

8. *Ibid.* book 1, 633–634.

9. Auguste Comte (1830–1842) Lesson 36, vol. I, p. 592.

10. E. Cassirer (1953) discusses the importance of the continuity of the substantial nature of qualities both for Aristotle and for the alchemical tradition.

11. Aristotle (n.d.) Book 1, Chapter 10.

12. G. E. Stahl (1730) Part I 'Structure of Matter', pp. 3–5.

13. The compositional interpretation of the chemical revolution has been emphasized by R. Siegfried and B. J. Dobbs (1968).

14. L.-B. Guyton de Morveau *et al.* (1787), p. 14.

15. E. B. Condillac (1780). Vol. 2, part 1, Chapter 2, p. 374.

16. For a discussion of this relationship in the context of an investigation into the sources of authority behind Lavoisier's nomenclature reform, see J. Simon (2002b).

17. P. Duhem (1902).

18. *Ibid*, translation by P. Needham, p. 5.

19. Post-Lavoisian chemistry is organized according to a logic that runs from simple to complex. A. Comte used this logic as the basis for his programme outlined in his *Course in Positive Philosophy* with the aim of realising a supremely rational chemistry: 'given the properties of all the simple bodies, to find those of all the compounds that they can form.' A. Comte (1830–42) Lesson 35, vol. I, p. 572.

CHAPTER 8

CHEMISTRY VERSUS PHYSICS

While it is evident that attempts to explain chemistry in terms of quantum theory could only appear following the development of quantum mechanics in the twentieth century, this type of reductionism has a much longer pedigree. In general, reductionism is the idea that a science, such as chemistry, can be reduced to a more "fundamental" science, in this case physics. As we shall see in what follows, the very conception of reductionism assumes a hierarchy in the sciences between the more fundamental sciences, usually physics and mathematics, and those that are seen as derivative or epiphenomenal, starting with chemistry, and passing via biology to the human sciences such as sociology or psychology. Thanks to the popular press and other media, we are familiar with the attempts to reduce human psychology to neuroscience, for example, although this kind of endeavour is recent when compared with the effort to reduce chemical interaction to the laws of physics. In 1669, Bernard Le Bovier de Fontenelle, the first and most famous lifetime secretary of the Royal Academy of Sciences in Paris, proposed the following comparison:

> Through its visible operations, chemistry resolves bodies into a certain number of crude tangible principles; salts, sulphurs, etc. while through its delicate speculations, physics acts on the principles as chemistry acts on bodies, resolving them into other even simpler principles, small bodies fashioned and moved in an infinite number of ways: this is the principal difference between physics and chemistry. [...] The spirit of chemistry is more confused, more dense; this spirit is more like mixts, where the principles are mixed together one with another, while the spirit of physics is clearer, simpler, less obstructed, and, finally, goes right to the origins of things, while the spirit of chemistry does not go to the end.[1]

While time has swept away Fontenelle's historical and scientific terms of reference, his comparison between the spirit of chemistry and that of physics remains relevant. Indeed, many scientists use similar arguments to justify the division of scientific or philosophical labour between physics and chemistry.[2] Thus, whereas chemistry is often seen to be aimed at providing the general principles governing material transformations, it is physics that is engaged in the quest for the ultimate constituents of nature and the fundamental laws governing their behaviour. From this classic perspective in the philosophy of science, chemists are faced with a choice: either they can accept the orthodox philosophical view and resign themselves to their subaltern status in a scientific world dominated by physics; or they can rebel and mount a heroic resistance against the imperial ambitions of their powerful neighbour. This dilemma will remain solidly in place in the philosophy of science for as long as we keep seeing such questions in terms of abstract epistemology relating to classic ontological issues. When science is seen as an essentially unified homogenous project aimed at solving well-defined ultimate ontological questions, then it is quite natural to oppose the valiant physicist — ready to assume the quest in its pure and complete form — to the narrow-minded empirical chemist who is unable to see the metaphysical wood for the practical, empirical trees. In this construction, it is clear that the physicist will always be in the forefront of scientific knowledge with chemists condemned to playing a second role. This is not, however, the only way of presenting the philosophical issues. Here, we want to suggest how chemists have come to define their own exigencies with respect to matter, as well as arriving at an understanding of their ethical responsibilities. In this respect, chemists have developed their own characteristic philosophical territory, quite independent from physicists and their familiar traditions and concerns.

To Each Science Its Ontology

Before exploring the issue of the philosophical differences between physics and chemistry any further, it is important to recall that the distinction between the two scientific disciplines is much more recent than the two philosophical traditions we are considering. As late as the eighteenth century, the term "physics" (from the Greek word "*phusis*" signifying nature) still referred to the whole field of natural sciences, with the exception of mathematics. Thus, chemistry was just a particular domain of physics.

While it may be true that the approaches adopted in the western alchemical tradition favoured elements and/or principles as the bearers of properties, this does not imply a rejection of atomist theories in this domain. Indeed, medieval European alchemy was moulded by the dominant scholastic culture, but it also reflected the thinking of those sidelined by the official church-sanctioned culture. Indeed, one of the reasons for the adoption of corpuscular and atomist notions by the alchemists was to defend themselves against the attacks mounted by scholastic philosophers. In Chapter 3, we saw the scholastic argument that the alchemists' gold only appeared to be gold but was not authentic because it did not contain the "substantial form" of gold that could only be produced by nature and not by artifice.[3] It was precisely to counter such attacks on the possibility of transmutation in the laboratory that certain thirteenth-century alchemists, notably pseudo-Djeber (the author of the *Summa perfectionis*), proposed a corpuscular theory of matter, although this was not explicitly linked to the ancient atomistic philosophy.[4] Keen to prove the authenticity of their artificial precious metals, the chemists argued that the chemical transformations took place *per minima*, at the level of the last term of analysis. Adopting this kind of corpuscular explanatory approach did not, however, exclude holding an elemental theory of matter. Indeed, as a series of recent studies of seventeenth-century chemistry have shown, these positions were seen as being wholly compatible.[5] Thus, Daniel Sennert, for example, explains a number of chemical operations in terms of the addition or the removal of atoms that possess their own characteristic chemical properties. We have to conclude, therefore, that neither the discontinuity nor the homogeneity of matter implied by atomism provides sufficient grounds to distinguish a philosophical view characteristic of the physicist from that of a chemist.

Historically, the concept of minima or corpuscles has provided a strategy for circumventing the Aristotelian distinction between matter and form. These indivisible minima were interpreted by the scholastics as possessing "substantial form", a mysterious property that the chemists succeeded gradually in doing away with. Indeed, one of the significant achievements of Robert Boyle, perhaps best remembered for his promotion of the "experimental philosophy", was to develop a vision of corpuscular philosophy that did not make any reference to the notion of substantial form. Boyle was a fervent experimenter, but his experiments were not limited to those concerning the vacuum generated by his famous air pump nor even to what we would today consider to be physics. He was a keen chemical experimenter as well, with his practical investigations covering the whole spectrum of seventeenth-century chemistry. This even included the transmutation of base metals into gold thanks to a small quantity of "philosophical mercury" he was able to procure.[6] Steeped in seventeenth-century chemical literature, he was still enough of a mechanical philosopher to dispense with the substantial form that was still present in the chemical writings of Daniel Sennert, a move that led to his launching an attack on the theory of elements or principles. This is the philosophical reorientation that lies behind the famous discussion of elements and principles presented by Boyle in his *Sceptical Chymist*. Here, Boyle's mouthpiece, Carneades, enumerates all the reasons for doubting the truly elementary nature of Aristotle's four elements — earth, water, air, and fire — before going on to do the same for Paracelsus's alternative threesome of salt, sulphur and mercury. While Descartes had argued around the purely theoretical question concerning what structure of matter could make the diverse properties of sensible bodies intelligible, Boyle bases his arguments on specific experiments that he himself has carried out. Thus, for example, Carneades explains how he has tried experimentally to separate gold into its elements, and has failed. Furthermore, in a rhetorical flourish, he offers to pay a reward to any chemist able to produce the three Paracelsian elements from this precious metal.

> I would fain see that fixt and noble Metal we call Gold separated into Salt, Sulphur and Mercury: and if any man will submit to a competent forfeiture in case of failing, I shall willingly in case of prosperous successe pay both for the Materials and the charges of such an Experiment. 'Tis not, after what I have try'd myself I dare peremptorily deny, that there may out of Gold be extracted a certain substance, which I cannot hinder Chymists from calling its Tincture or Sulphure; and which leaves the remaining Body depriv'd of its wonted colour. Nor am I sure, that there cannot be drawn out of the same Metal a real quick and running Mercury. But for

the Salt of Gold, I never could either see it, or be satisfied that there was ever such a thing separated, in *rerum natura*, by the relation of any credible eye witnesse.[7]

According to Boyle, there is no reason to admit that a given body is composed of the traditional selection of elements (either Aristotle's four elements, or the *tria prima* of Paracelsus) because the decomposition of almost any complex substance provides other equally plausible candidates for the status of element. Furthermore, nothing proves that any of these prospective elements that result from analysis actually existed in the decomposed substance prior to its decomposition. Instead, Boyle favours the theory that all bodies are formed out of a single "catholic" matter and suggests that while necessary, even a mechanistic explanation of their qualities based on the form and motion of the constituent corpuscles may not be sufficient. The critical part of Boyle's book is based on a definition of the element as a simple body, or what remains after analysis. Indeed, this preliminary definition that serves as the measure against which he can deploy his scepticism, has earned Boyle the honour of being considered the founder of modern chemistry in the eyes of several historians.[8] Nevertheless, this historical claim is doubly unjust. First, because Boyle acknowledges that this definition of an element already exists among other chemists. Second, because he does not believe that any material body answers to this definition of an element.

Even Boyle, therefore, plays on both sides of any alleged divide between atomic theory and principles or elements. Thus, we can conclude that while tempting, it is ultimately too superficial to try to distinguish chemistry from physics along the line of elements versus atoms, between a philosophy of matter characterized by inherent principles or a vision based on homogeneous but discontinuous matter. First, as we noted in Chapter 7, the clash between these two traditions preceded the emergence of the modern disciplines of physics and chemistry. Second, and more significantly, chemists have always been confronted by a multiplicity of interpretations for chemical phenomena, and a large number of them have not hesitated to wed corpuscular doctrine with that of elements or principles.

The critical part of Boyle's book was aimed at those he referred to as "vulgar chymists," meaning the professors occupying the chairs of chemistry created in the medical faculties or those chemists teaching private courses for apothecaries and doctors. Many of these teachers both from within the faculties and outside supplemented their courses with the publication of a textbook. Thus, while these texts were principally composed of practical instructions for preparing medicines, they also presented a scientific framework for understanding chemical change organized around these kinds of principles.[9] It is important to understand that the term "principle" covered both the material constituents of bodies that could be obtained by analysis (or "separation" of the body using fire) and the immaterial intangible, irresolvable constituents, also known as "spirits". In both cases, the principles were seen to be bearers of certain qualities, "virtues", or properties that they could confer on the mixts they entered into. When Fontenelle mentions the chemist's "crude tangible principles" in the passage cited at the beginning of this chapter, he is referring to the *tria prima* of Paracelsus and his followers: salt, sulphur and mercury. Each one was considered to bear specific properties, salt solidity, sulphur inflammability, and mercury fluidity. Despite differences of opinion concerning the nature and the number of these principles in the chemical literature of the seventeenth century, this central concept allowed the authors of these treatises to present a novel and coherent vision of matter. Thus, in his *Elements of Chemistry* of 1610, Jean Béguin could not only present chemistry as an auxiliary medical science but also as the bearer of a distinctive vision of matter.

> I want the physicists and doctors to understand that chemists are not attacking them when they use alternative principles. Just as Aristotle and the rest of philosophy teaches us that two arts or sciences can have the same matter, or the same material body as their object but not consider them by means of the same specific intrinsic principles or according to the same formalism, and if they also affirm that chemistry is a different art from physics or medicine, then they have to agree with us that chemistry should dispose of different specific intrinsic principles formally constitutive of its object. In order to illustrate this theory, I would say that the physicist, the doctor and the chemist can very well treat the same body under consideration in various ways and according to diverse principles. The physicist will consider it as natural, and capable of motion or rest according to the

constitutive principles of the natural body, in terms of their nature, that is matter and form, because this is how it is constituted as an object of physics. The doctor will consider the same body in terms of its capacity to receive or cause health, examining it in terms of the four primary qualities, cold, hot, dry, and wet that constitute the body's temperament, which is responsible for health or sickness. The chemist, likewise, will consider the object in his own way, in terms of its ability to be resolved or to coagulate and the diverse virtues in its interior that can be made manifest by the art, rendering it more useful. Thus, in so far as mercury, sulphur, and salt are the principles that render mixt bodies soluble or prone to coagulate — being the roots of the body's internal virtues — or that they are veritable chemical substances, that is to say the principles that support and give substance to all the virtues and accidents of the compound, the chemist should proceed by means of these three principles in all his examinations, theories and operations.[10]

Following the demise of his four-element theory, the lesson that Béguin draws from Aristotle is that to each science or art its own specific perspective on matter. The principles of chemistry "give substance" to matter, in both senses of the term: they nourish and give physical substance to the body, while at the same time reifying the associated properties. However many principles one might accept, each one has its own specificity and function. These principles are not, however, intended to explain the properties of a given compound, but rather to throw light on what happens when, for example, an acid dissolves a metal. The principles provide the means for describing and interpreting just these sorts of chemical operation that make up the body of this kind of treatise on chemistry.[11]

The Ultimate Quest

In the passage cited above, Béguin is concerned with the ambiguity of the term "principle", pointing out that it can be used to designate both the basis of knowledge (epistemological sense) and the universal constituents of natural bodies (ontological sense). Fontenelle puts the epistemological and the ontological in relation in a different way. In his discussion of chemistry and physics, he argues on the basis of an epistemological principle that the specific properties of material bodies need to be reduced to the smallest possible number of discrete ontological principles. The article "chymistry" written for Diderot's *Encyclopedia* by Gabriel-François Venel, a doctor and chemist from Montpellier, is presented as a reply to Fontenelle's depreciative reflections on chemistry, which Venel interprets as an expression of Fontenelle's prejudices concerning the field. He translates this feeling into a vigorous assault on the arrogance of the physicists, which presents interesting parallels with recent defences mounted by chemists against the hegemony of quantum mechanics as the only legitimate explanation of chemical phenomena. According to Venel, chemists are not so much concerned with the ultimate questions of the nature of matter as with questions like "What gives *acqua regia* its ability to dissolve gold?" Their attention is focused on the specific properties manifested by substances and their transformations rather than questions concerning the structure of matter. For Venel, "no body is just matter".[12] Indeed, the qualities responsible for the chemical properties of a substance, like acidity, flammability, etc., are incorporated in the principles that then carry them from one mixt to another. This theory allows chemists to explain the circulation of properties while retaining the fundamental notion of the conservation of matter. Thus, Venel is proposing a substantival interpretation of the particular properties manifested by any given body, where Fontenelle clearly favours the idea of a unique uniform matter that constitutes the substance of all things, and gives rise to its phenomenal texture.

The physicist's spirit, which Fontenelle finds so seductive and Venel so irritating, is nothing other than the Cartesian theory of uniform homogeneous matter; pure extension possessing length, width and breadth, and put into motion by God. Nevertheless, this conception of matter, which is tied to a particular metaphysics illustrated by the move from the first to the second part of Descartes' *Principles of Philosophy*, while it allows the philosopher to conceptualize physics in terms of geometry and to present the laws of motion, is not intended to offer any direct explanation of the world of phenomenal appearances, let alone give an account of chemical reactions.

Having said this, however, Descartes does try to provide a total philosophy of nature — including chemical phenomena as well as the formation of the world — in the two final parts of his *Principles of Philosophy*, albeit without reference to the qualities of bodies. Indeed, his goal is to explain all the phenomenal properties, which the chemist usually credits to the principles, in terms of the diversity of the extended form and motion of the homogenous matter that makes up the world. In the fourth part of the *Principles*, Descartes presents a hypothetical picture of the constitution of chemical reactants in line with the idea that the uniform matter created by God was differentiated by internal vortices.[13] The friction generated by the movement of these vortices has over time divided this matter into three species: luminous matter that makes up the sun and the stars; transparent matter composed of "round and very small" particles that form the sky; and the matter of opaque bodies, larger irregular particles that make up the surface of the earth.[14] The particles of Descartes' three "elements" only differ from one another with respect to their size and form, meaning that all the stuff of the world — earth, water, fire and air — is composed of three dispositions of the same matter. Thus, the Aristotelian elements are neither considered to be the original constituents of matter nor are they truly elementary or irreducible. Descartes conceives the elements, like all other chemical substances, as the products of a history of the material world, the products of geological chemistry as it were, and for him, being simply conformations of a unique homogeneous matter, the elements can lay no claim to any special ontological status.[15]

Thus, we are left with the impression that the opposition between Venel's defence of elements or principles and the Cartesian vision of homogeneous matter varying only in form and size is just a transposition of the opposition between atomism and Aristotelian matter theory that we examined in the previous chapter. Matter is either intrinsically differentiated or it is homogeneous and uniform; the former approach posits principles or elements as the irreducible bearers of properties, while the latter considers the same properties to be secondary qualities or epiphenomena that arise due to the form and state of motion of the matter in question.

No Matter Without Qualities

We will now attempt to identify the veritable points of cleavage that separate the "spirit of physics" from the "spirit of chemistry". The divergence does not concern atomism as such, understood as a theory that postulates the existence of a vacuum and discrete units of matter. Indeed, the discontinuity of matter implied by atomism fits so well with the explanations of the phenomena of chemical combination that the laws of definite and multiple proportions served as a key argument in favour of adopting chemical atomism in the nineteenth century. No, the point of dissent is rather the reduction of qualities to other parameters considered fundamental such as figure or shape and motion.[16] According to Descartes, one cannot arrive at the true principles of nature based on experience because the senses do not communicate any reliable information about the fundamental make-up of the world. The true principles arise from considering "division, shape and motion". The real structure of matter remains inaccessible to the senses, even when their power is amplified by technological aids like the microscope. Thus, colours, smells and other sensible properties are the result of the interaction between the shape of the constituents of a substance and the experiencing subject. In other words, Descartes adopts the scholastic distinction between primary qualities — belonging to the essence of a body — and secondary qualities, which depend on the senses. This distinction, however, serves to disqualify sensible matter in the most literal sense of the word "disqualify"; that is to say depriving it of its qualities. At the same time, it discredits all chemical theory that deals with the properties of matter. According to the Cartesian vision, the essential primary qualities are limited to figure and motion, and are perceived by the intellect, while any quality perceived by the senses (colour, smell, etc.) is derived from and fully determined by these primary qualities. As Descartes puts it quite clearly in his *Regulae at directionem ingenii* (*Rules for the Direction of the Mind*), "it is certain that the infinite number of figures is sufficient for expressing all the differences that exist between sensible things."[17]

The distinction we arrive at is between "substantival realism" and mechanistic corpuscularism, with, on the one hand, qualities endowed with material reality, and, on the other, all sensible qualities derived from the form and motion of the constituent corpuscles. For those corpuscularists who believe that form is a sufficient principle for understanding the natural world, the chemist's qualitative theories appear to be overly clumsy and approximate. Nevertheless, the chemists do not go "right to the end" because of their attachment to sensible qualities. Gaston Bachelard revived, or rather continued this tradition in his well known book, *The Formation of the Scientific Spirit*, where he presents sensible qualities as obstacles to be overcome. In their cognitive enterprise, chemists are impeded by an interest in secondary qualities that diverts them from uncovering the mechanical basis of such properties that would in turn render them accessible to a geometric treatment.

It is just this kind of reductionist ambition on the part of the physicists that pushes the chemist into philosophical revolt. Today's rebellious chemist (along with many biologists and even physicists) rejects the hierarchical model of science that lies behind this reductionism. This model enthrones mathematical physics at the top of a disciplinary pyramid in which sociology is to be reduced to biology, biology to chemistry, and chemistry to physics. Nevertheless, even before this modern disciplinary understanding of reductionism was in place, chemists asserted their own system of scientific values in face of the hegemonic explanatory ambition of mechanical philosophy. It was not, after all, difficult to poke fun at the fantasies or "fables" generated in the quest to explain a substance's chemical properties — corrosiveness, bitterness, or mildness, for example — in terms of the shape and motion of its constituent particles. Perhaps the best known example of this project in seventeenth-century chemistry is to be found in Nicolas Lemery's *Cours de Chymie*. Here, for example, he offers a model corpuscular explanation of why *aqua regia* (a mixture of nitric and sulphuric acid) precipitates the gold it holds in solution upon the addition of an alkali. The explanation was based on the idea that acidic corpuscles were pointed, with their pointed form giving rise to the acidic taste as a result of the corpuscles pricking the tongue. The action of an alkali on an acid was always to neutralize the acidity, and so Lemery concluded that the alkali must in some way break the points off the acidic corpuscles. Finally, he used the idea that the gold particles were porous to explain the dissolution of gold in terms of the points of the acid lodging in the pores. Thus, all the elements are in place for Lemery's mechanist explanation of the dissolution of gold in *aqua regia* (with the phenomenon of the saturation of the solution thrown in) and its subsequent precipitation by alkalis.

I suppose that when the *aqua regia* acted on the gold in such a way that it dissolved the gold, the points which are responsible for the acid's strength were stuck into the particles of gold. But, because these little bodies are very hard, and consequently difficult to penetrate, the points only enter superficially, although far enough to suspend the particles of gold and to prevent them from precipitation; that is why, add as much extra gold as you will, when each of these points has taken up what it can support, it will not dissolve a grain more of it; it is also this suspension that renders the particles of gold imperceptible. But if you add some body, that thanks to its motion and figure can, by this shock, shake-up the acids enough to break them, the particles of gold, being free, will precipitate due to their own weight: this, I claim, is what the oil of tartar and the volatile spirits of alkali do.[18]

The multitude of different corpuscles — smooth and round, jagged, or hooked — proposed by Lemery provoked the satirical scorn of a number of chemists. For Venel, these speculations only serve to reveal the insufficiency of the mechanists' axiomatic approach as well as their lack of empirical experience. While he admits that physicists might be competent to treat certain types of compounds, known as "aggregates", Venel reserves the treatment of "mixts" for the chemist. He thus reinvigorates Stahl's distinction between aggregates which are simply the result of a mechanical union or juxtaposition of different types of substance, and mixts which involve the chemical transformation of the substances involved. The aggregates, whether understood in terms of the interlacing of corpuscles or the play of Newtonian attractions, involve the general properties of bodies such as mass and motion, which are legitimately handled by mechanics, a science that covers the behaviour of homogeneous matter. The formation of a mixt, however, implies the transformation of the reactants, giving rise to qualitative change, and a fundamental shift in the identity of the matter. The process of "mixtion" or reaction creates new homogeneous bodies out of heterogeneous elements, and no explanation in terms of the simple physical juxtaposition of particles can conceivably be adequate to the phenomena involved.

Thus, Stahl's distinction between aggregates and mixts served to draw a frontier between the legitimate domain of physics and that of chemistry. Not only does this approach overthrow the hierarchy in which physics offers the ultimate explanations, it effectively inverts it by suggesting that by trying to explain the phenomenon of the mixt, chemistry addresses much more complex and mysterious questions than those that aggregates pose for physics. Chemistry's object is the analysis or *diacresis* of mixts. In order to understand the nature of a mixt, the mechanical interaction of the combined substances was clearly insufficient for the chemist of the time. An approach that advocated endowing the original substances or their constitutive principles with all the sensible qualities of the mixt would inevitably lead to counter-intuitive results as well as a radical inflation of the hidden or potential properties of matter. Furthermore, as we have already seen, the mechanical philosophy proposed that substances did not possess any of the sensible qualities normally ascribed to them. Thus, the particles of gold that constitute a metallic, yellow, ductile sheet of gold are neither yellow, nor metallic, nor ductile.

We should not let Fontenelle's rhetoric fool us into thinking that it was only the physicists who had something to say about the composition of matter either. By the eighteenth century, chemists were working with a complex and increasingly well-ordered hierarchy of analytical products, particularly in the animal and vegetable kingdoms of nature. Here, the chemists distinguished between the ultimate principles that we have been discussing above and the "immediate principles" of analysis. While opinions varied concerning the number and precise nature of the ultimate principles, there was a general consensus that they were inaccessible, and although analysis could give the chemist an idea of the constitution, it would not yield the principles in their pure form. Indeed, like other "essences", their presence can be known by their effects and not through their isolation. The immediate principles, however, were substances such as the oils, or the various fats that constituted homogeneous isolable materials entering into the complex composition of living plants and animals. Over the course of the seventeenth and eighteenth centuries, in particular with the development of increasingly sophisticated solvent extraction, chemists had identified a wide range of these intermediate principles.[19] This kind of multi-level approach to the analysis of matter shows that the chemist was not afraid of addressing the most difficult questions concerning the nature of the material world, and serves to remind us that these in fact involve treating nature in its complexity rather than laying aside these issues in order to try to arrive at the simplest parts. For Venel, unlike physicists, chemists try to understand material substances in their complexity and their individuality, and instead of resting at the surface, they try to understand the very depths of the natural world. While clarity, distinction, and mathematical abstraction may have historically been the pride and joy of the physicist, one can argue that their excessive valorization may lead the scientist to leave the most significant parts of nature to one side. According to Venel, physics exorcizes its own fear of the unknown by reducing all physical properties to questions of geometry, as though truth was identical to or at least co-extensive with a mathematical model of intelligibility. Chemists, on the other hand, have the courage to recognize the diversity of the principles that enter into mixts and to admit that their knowledge has its limits. Thus, the chemists' view that the ultimate principles are unknowable signifies both their inability to provide the ultimate answers, and their awareness of this limitation. While this argument in its details applies to the image of chemistry promoted in Diderot's *Encyclopédie* more than two centuries ago, it nevertheless provides valuable clues for understanding the philosophy of modern chemistry.

Thus, we can see how chemists defined, even though only implicitly, their own characteristic ontological requisites precisely by refusing to follow the physicists on the single-minded quest for the ultimate constituents of matter. While the physicists were forming hypotheses about the minimal units that structured phenomenal matter — be it continuous or discontinuous — along with the ultimate causes of the behaviour of matter, chemists were content to retain their elements cum principles or to adopt the physicists' atoms cum corpuscles or even, as we have seen, to try to combine the two. Chemists are not opposed to proposing any theory of matter, rather they object to the idea that one can reduce the diverse specific qualities of bodies to motion and form, or any such simple geometrical parameters. The chemist's principles are not supposed to represent an alternative philosophy of matter intended to compete with atomism, but are instead retained for their value in interpreting chemical phenomena.

The prime philosophical imperative for chemists seems to have been, and still remains: do not disqualify matter in its materiality. This is not only because the sensible qualities or properties of matter are considered to be essential and not secondary effects of some deeper reality, but also because matter has the potential to act in the world; every substance with its own specific dispositions.

No Matter Without Agency

While many physicists in the eighteenth and nineteenth centuries saw fit to adopt the corpuscular or atomic theory as a way to approach a number of domains, the chemists found it largely inadequate. The central problem for the chemist with respect to the corpuscular philosophy was the dearth of properties attributed to matter that would allow it to act on other matter. It seemed wholly implausible that differently shaped small bodies getting caught up or bouncing off one another could explain the range of chemical phenomena that one encounters in and out of the laboratory. The response was to adopt other means of action to augment the inherent potential agency of matter. Thus, two strategies for boosting the agency of matter established themselves in European chemistry: the Newtonian approach that dominated the scene in France and the dynamist one that enjoyed more success in the German states.

One of the best-known stories in the history of physics is that of Newton's rejection of the Cartesian mechanical philosophy by his insistence on action at a distance as an irreducible natural phenomenon. Furthermore, according to Newton, gravity was not the only principle which inhered in bodies.

[I]t seems probable to me, that God in the Beginning form'd Matter in solid, massy, hard, impenetrable moveable Particles, of such sizes and Figures, and with such other Properties, and in such Proportion to Space, as most conduced to the End for which he form'd them; and that these primitive Particles being Solids, are incomparably harder than any porous Bodies compounded of them; even so very hard, as never to wear or break in pieces; no ordinary Power being able to divide what God himself made one in the first Creation. [...] It seems to me, farther, that these Particles have not only a *Vis inertiæ*, accompanied with such passive Laws of Motion as naturally result from that Force, but also that they are moved by certain active Principles, such as that of Gravity, and that which causes Fermentation, and the Cohesion of Bodies. These Principles I consider, not as occult Qualities, supposed to result from the specifick Forms of Things, but as general Laws of Nature, by which the Things themselves are form'd; their Truth appearing to us by Phænomena, though their Causes be not yet discover'd.[20]

In his famous Query 31 of the *Opticks*, Newton supposes that a single force of attraction that obeys an inverse square law with respect to distance, but, unlike gravitation, diminishes as the size of the particle increases, might be able to explain the violent reactions between acids and bases.[21] This proposition served to re-legitimate chemists' talk of the "tendencies", "virtues", and "powers" of their reagents. The key to opening up a line of communication between the macroscopic and the microscopic, between the sensible world in which we live and the hidden world of particles and forces is to endow matter with "virtues" in the double sense of forces (the meaning of the Latin root, *virtus*) and virtuality or potentiality. The "clear and distinct" that served as the fundamental guide in Descartes's epistemology need to be set aside in favour of a sprinkling of "occult" action. According to Newton's proposition, one single force involving action at a distance may be enough to explain the chemical mystery of elective affinity. Indeed, there is a historical question as to the possible origin of Newton's concept of attraction in chemistry itself. While the schoolroom myth is that the law of gravitation came to Newton upon seeing an apple fall to the ground, the more scholarly version maintains that gravitation was the outcome of his interpretation of empirical astronomical data. Nevertheless, we now know that Newton had a lifelong interest in chemistry, and his painstaking quantitative experiments in this discipline were aimed at understanding the mechanisms behind chemical attraction. Indeed, certain eighteenth-century chemists, including Rouelle, were convinced that Newton's concept of attraction was chemical in origin.[22]

Whatever the origin of his idea of universal attraction, Newton's Query 31 furnished chemists with a research programme that lasted for almost a century. Determining the proposed law of attraction promised to provide a means for making the individual "predilections" of each substance commensurable, allowing future generations not only to compile more accurate affinity tables, but also to predict the outcomes of unknown reactions. Thus, the notion of affinity–attraction as a constant elective property of substances is a commonplace in eighteenth-century chemistry. Nor does the use of this term necessarily imply a reference to Newton, as it could be interpreted in purely phenomenological terms as a relationship between given substances, that is to say their tendency to unite chemically with one another. The idea of such a short-range force of attraction allowed chemists to explain a range of mysterious phenomena. Thus, for example, it could be used to explain the cohesion of homogenous bodies, as in the case of a simple substance like a gold brick composed of identical constituent parts. Alternatively, the same force could explain "affinity" understood as the cause of the combination between the components of a heterogeneous body. In turn, the affinity of aggregation was differentiated from the affinity of composition, with only the latter implying a qualitative change in the combined components. The ultimate goal was to use the disposition of bodies to supplant the occult concept of attraction, as the hypothesis of a simple relationship between two substances seemed more acceptable. While it was mobilized to justify increasingly complex tables of reactions, affinity remained a qualitative rather than a quantitative concept. Overall, affinity–attraction underwrote a coherent, explanatory and predictive theoretical system that allowed chemistry to assert its independence with respect to physics.

The production of affinity tables has a long history in chemistry. The best known and most widely reproduced of these essential aids to the practicing chemist was drawn up by Geoffroy in the early eighteenth century, and was reprinted in a slightly extended version in Diderot's *Encyclopédie* some forty years later (see Figure 4). Each column represents a series of potential chemical combinations between the substance or radical indicated at the head of the column and the others listed in a specific order underneath it. A substance that entered into chemical combination with the one at the head of the column could, in principle, be displaced from this combination by any substance featuring above it in that column. These tables were continually revised and improved in the following decades to take account of different reaction conditions and so represented the compilation of an enormous quantity of empirical data concerning the reactions between simple and complex bodies. Nevertheless, the general form of these tables and the observable hierarchy among the components in terms of reactivity give an overall qualitative measure of their various affinities. This painstaking empirical chemistry is far removed from the "Newtonian dream" that led a handful of chemists and physicists at the end of the eighteenth century to try to reduce all chemical phenomena to Newton's proposed short-range attraction.[23] At base, this dream of chemistry determined by well defined mathematical laws was one shared by Claude-Louis Berthollet. While testing the system of affinity as it had been elaborated in the affinity tables, he launched an attack on the numerous weaknesses of this system. Indeed, he challenged the theory that each substance had a constant affinity, as this principle was unable to explain the course of many well known chemical reactions. The tendency for two substances to form a new combination, which he dubbed "chemical action", does not only depend on the reciprocal affinities of the substances in question, but also depends on their respective proportions in the reaction medium. On these grounds, Berthollet denounced the affinity tables, which, he believed, expressed differences in solubility and volatility more than any real differences in affinity. Having clarified the parameters that entered into chemical reactions, Berthollet, helped by Laplace and their disciples at the Society of Arcueil, undertook the project of making chemistry mathematical on the model of mechanics, hoping to base the science on one single mathematical law from which all the reactions could be deduced. Nevertheless, Berthollet was unable to unite chemistry and mechanics on this model, and his concept of chemical action remained marginalized for decades.[24]

Newtonianism also served to revitalize corpuscularism by endowing matter with forces denied to it by Cartesian philosophy. Matter was empowered by the addition of action-at-a-distance to the fundamentalist ontology of form and motion. But there is an alternative vision of empowered matter that avoids the mechanist's vision of uniform matter and corpuscles altogether. It is possible to conceive nature as a collection of opposed forces that engender the phenomenal properties of bodies through their polarity. This was the vision of nature incorporated into the *Naturphilosophie*, the philosophy that dominated the universities of the German states at the close of the eighteenth century. In a context where the sciences were taught in the faculties of philosophy, chemistry lent itself to a "dynamist" interpretation. This tradition, championed by Richter, Klaproth, Ritter, Kastner, Link, and Schuster, as well as Oersted from Denmark and Winterl from Hungary, was characterized by a strong association between experimental chemistry and speculative philosophy that formed the basis for a discussion of the foundations of the positive sciences. In this context, it was quite natural that both Hegel and Schelling should take such a keen interest in the evolution of positive science.[25]

The basis for the elaboration of this dynamist chemistry was provided by the sensible qualities of chemical substances. The explanation supposed that matter was animated by forces — magnetic, electric, chemical — with opposite polarities, and that it was these polarities that determined the phenomenal qualities of the matter. This treatment of qualities was inspired by Kant's notion of intensive quantity that he proposed in his *Metaphysical Foundations of Natural Science* from 1786. In contrast to form and motion, intensive quantities are not amenable to mathematical representation, but can nevertheless be handled quantitatively as non-additive quantities. The science of stoichiometry developed by Jeremias Richter (1762–1807) (one of Kant's students) illustrates this approach nicely. Wanting to introduce a mathematical approach to experimental chemistry, Richter quantified the properties of being acidic or basic by placing them on a scale, thereby allowing him to determine the proportions of the reactants involved in the formation of salts, leading to his proposal of the law of neutralization.

Although developed a generation later, Jons Jakob Berzelius's electrochemistry can be considered as part of this movement. Berzelius conceived chemical affinity as the result of the play of electric polarities, and at the same time explained the definite proportions observed in chemical combinations in terms of the combination of the corresponding atoms of each element. Hegel, on the other hand, who dedicated a part of his *Encyclopedia of the Positive Sciences* to chemistry, rejected any association of the stoichiometric proportions of compounds with the idea of atoms, as he rejected the material nature of the element itself. Hegel believed that the multiplicity of such elements resulted from diverse combinations of chemical polarity and not from any fundamental material diversity, making it logical that he reject any attempt to reduce chemical phenomena to those of physics. This dynamist tradition with its characteristic mix of chemistry and philosophy would fall victim to the success of Lavoisier's new chemistry in the nineteenth century. In the wake of this revolution, chemists were quick to condemn the *Naturphilosophie* that had reigned in the German universities as obscurantist and unscientific, with Liebig's description of this approach as the "black death" of German chemistry being only one attack among many. As for Hegel, his contribution to chemistry has been largely forgotten, with his notion of dialectics bearing its fruit in political and economic philosophy rather than the philosophy of science.

At the risk of falling victim to oversimplification, we have painted a historical fresco of the evolution of chemistry using broad brush strokes. This historical overview nevertheless serves to illustrate how chemists have been driven to develop their own philosophy of matter. Initially, their priority was interpreting and predicting the individual behaviour of a given substance in a well defined context, a task that is poorly served by the abstract concept of matter. The chemists' principal aim is not to understand what lies behind the vivid and sometimes explosive transformations of the multitude of reactants that they handle. Cartesian substance, characterized by its extended form — length, breadth and width — and its motion was elaborated by physicists in a quest to penetrate to the essence of matter. This quest was not shared by the chemist, who has been either unwilling or simply incapable of stripping matter down to this bare ontology, or, more literally, of disqualifying matter in this way. Matter without qualities, matter that is necessarily informed by something else, matter whose phenomenal place in the world depends solely on its form and motion is simply inadequate for dealing with the rich active world encountered in the chemist's laboratory. Chemistry's theatre cannot dispense with its players, who are individuals with the capacity to act and react, and whose existences are interwoven in a complex web of relationships. Thus, chemistry's drama is inevitably richer than the reductive dream that has been characteristic of the history of physics.

References

1. B. Fontenelle, *Histoire de l'Académie royale des sciences*, vol. 1, comments on the year 1669, cited in H. Metzger (1923), pp. 266–268.
2. Fontenelle's comparison was developed in a specific historical context that involved a debate between two interpretations of chemical transformations. Gaston Du Clos's interpretation in terms of principles in his 'Dissertation on natural mixts' was opposed to the corpuscular interpretation proposed by René Descartes and Robert Boyle.
3. C. Lüthy, J. E. Murdoch, and W. R. Newman eds (2001).
4. W. Newman (1991).
5. A. Clericuzio (2000) and W. R. Newman (1996) and (2006).
6. See, L. Principe (1998) and W. R. Newman (1996).
7. R. Boyle, (1661), pp. 174–175.
8. Ibid. A résumé of Boyle's arguments can be found in H. Metzger (1923), pp. 251–266. More recently, Lawrence Principe has offered a new interpretation of the Sceptical Chymist in the light of Boyle's alchemical practices. See, L. Principe (1998), Chapter 2.
9. The great majority of these courses adopted Paracelsus's *tria prima*, even though they often complemented them with the 'passive' elements, earth and water. For an examination of this tradition of chemistry courses in France, see H. Metzger (1923). M. Bougard (1999) has identified more than 150 such treatises. For an argument concerning the central role of such courses in the constitution of the discipline of chemistry, see O. Hannaway (1975).

10. Jean Béguin, *Éléments de chimie*, Paris, 1610, pp. 27–28, cited in H. Metzger (1923), pp. 38–39.

11. Following the introduction and a presentation of the experimental apparatus that a chemist needs, Béguin's book is made up of a collection of recipes for making medicines with descriptions of their therapeutic virtues and their common uses. For more about this tradition of pharmaceutical chemistry, see J. Simon (2005), Chapter 5.

12. Venel, in the article 'principles' in the *Encyclopédie*: 'nul corps n'est de la matière'. Diderot, Denis, and Jean D'Alembert, eds (1751–1765).

13. Thus, water is composed of small flattened corpuscles that are able to slide across one another, while the corpuscles of the earth wind around each other like the branches of a tree. The salts are composed of particles that have become thin and pointed by a hammering action; sulphur, or oily matter, is formed from wrinkled softer corpuscles that stick to one another due to their ragged form. Descartes concludes this corpuscular hypothesizing in the following terms: 'Here, I have explained three sorts of body, which, it seems to me, are close to those the chymists usually take to be their three principles & that they call salt, sulphur, and mercury.' R. Descartes (1647), Part 2, articles 46 and 47.

14. Ibid article 52. The second element is composed of smaller, faster moving spherical corpuscles, and the third element is the most subtle of all. This third element, the 'dust' formed from uniform matter, is constituted by the smallest corpuscles and fills the interstices left empty by the two others.

15. R. Descartes, (1647) Part 4, article 63, Version Adam-Tannery IX-2, p. 235. See B. Joly (2000).

16. Thus, Descartes is at the same time opposed to atomism and hostile to the principles of chemists. In fact, he considers Democritus's atomism to be based on two principles – the atoms and the vacuum – that fall victim to the same criticism he levels at Aristotle's four elements and the three Paracelsian principles. See the preface to the French edition of R. Descartes (1647), pp. 7–8.

17. R. Descartes (1628–1629), Règle XII.

18. N. Lemery (1675), pp. 43–44.

19. See, F. L. Holmes (1971)

20. I. Newton (1730), pp. 400–401.

21. Idem, p. 375 ff.

22. In commenting the Query 31, Rouelle declared that 'Newton explains dissolution by attraction; the idea of which undoubtedly came from his chemical experimentation.'
23. M. Crosland (1967).
24. P. Grapi (2001) and F. L. Holmes (1962).
25. E. Renault (2003).

CHAPTER 9

ATOMS OR ELEMENTS

Today, we are surrounded by artificial chemical substances, ranging in sophistication from domestic bleach to the modern polymers that encase our mp3 players or form the resistant kitchen surfaces that wipe clean with a brush of an equally artificial "sponge". Nevertheless, despite the omnipresence of these chemical products, no particular substance symbolises chemistry as effectively as a simple chart displaying all the elements known to modern science presented in a specific order. Indeed, just the outline of the periodic table with its characteristic block layout is enough to evoke an association with the discipline of chemistry in the minds of scientists and laymen alike. It was the discovery of the periodic repetition of similar properties exhibited by a series of elements of increasing atomic weight that first allowed the organization of all the known elements in an ordered structure in the nineteenth century. While today's chemistry student learns the explanation of the periodic table in terms of the electrons filling the orbitals around a nucleus, this theoretical justification was completely absent when Mendeleev first proposed the periodic relationship. Indeed, for this Russian chemist working in the 1860s the existence of these series of elementary octets was an entirely empirical proposition. Mendeleev was, however, happy to go beyond simple empirical observation and even left empty spaces in his table; spaces that were to be occupied by elements yet to be discovered. Thanks to his table, Mendeleev was also able to estimate the atomic weight of these currently unknown elements.

Despite the proliferation of the ways of representing the chemical periods, the simple block structure dominates schoolrooms and boardrooms alike, serving along with the test-tube and beaker as a universal symbol of chemistry. In constructing the original periodic table which dates from 1869, Mendeleev opted for the element-principle approach, rather than atomism, adopting one of the essential attributes of the ancient Aristotelian elemental approach, pluralism. In Mendeleev's case, however, this pluralism related qualitative differences to a quantitative variation in the value of the atomic weight.

Mendeleev's Wager

All his life, therefore, Mendeleev vigorously defended the credo of the irreducible plurality of elements in the face of numerous opponents who sought to reduce this plurality to one single primordial element, usually the first one in the periodic table, hydrogen.[1] Nor was Mendeleev alone in trying to put some order into the variety of elements, with others joining him in the search for a taxonomy that could aid in the understanding and manipulation of the profusion of elements including both the ancient ones and those that had only recently been discovered. The known properties that could enter into such a classification were not limited to the atomic weight (the value now known as the atomic mass), but featured a range of chemical properties such as combining power or valency, degrees of oxidation, reactivity, and physical state. While these are all phenomenological or experimentally determined attributes, they nevertheless, like any empirical data, depend on specific hypotheses or some kind of theory.[2]

In the case of the periodic table, however, theoretical considerations already entered into the definition of the object to be classed — the chemical element. At the end of the eighteenth century, Lavoisier put considerations of the ultimate constitution of matter to one side and instead formalized a pragmatic definition of the element that conformed with the dominant use of the term "simple bodies". Indeed, Lavoisier's definition of the chemical element in his *Elements of Chemistry*, at once more categorical and more useful than Boyle's sceptical reflections, has assumed the status of a landmark in the history of chemistry.

[I]f, by the term *elements*, we mean to express those simple and indivisible atoms of which matter is composed, it is extremely probable we know nothing at all about them; but, if we apply the term *elements*, or *principles of bodies*, to express our idea of the last point which analysis is capable of reaching, we must admit, as elements, all the substances into which we are capable, by any means, to reduce bodies by decomposition. Not that we are entitled to affirm, that these substances we consider as simple may not be compounded of two, or even of a greater number of principles; but, since these principles cannot be separated, or rather since we have not hitherto discovered the means of separating them, they act with regard to us as simple substances, and we ought never to suppose them compounded until experiment and observation has proved them to be so.[3]

Lavoisier defines the element as the last term of analysis, but has nothing to say about the relationship between the substance that the chemist might be able to obtain and its (homogeneous, similar) constituent parts (see Lavoisier's list of simple substances reproduced in Figure 9). Mendeleev, by contrast, raises this issue right at the beginning of the article in which he presented the first outline of his table. Here, he makes a distinction between a chemical element and a simple body.

A simple body is something material, a metal or a metalloid, endowed with physical and chemical properties. The idea that corresponds with the expression simple body is that of the molecule [...]. By contrast, we need to reserve the name element to characterize the material particles that constitute the simple bodies and compounds and that determine the manner in which they behave in terms of their physical or chemical properties. The word element should summon up the idea of the atom.[4]

Thus, the simple body is the "element" in its gross substantial form, Lavoisier's last term of analysis, but Mendeleev has transposed the term element onto the unit whose aggregation constitutes this simple phenomenal body. Such a distinction was superfluous in the context of a chemistry based on analysis, essentially preoccupied by the interplay between simple and compound substances on the phenomenal level. Nevertheless, the distinction becomes particularly pertinent when the goal is to manage an increasingly rich, complex and apparently irregular collection of diverse substances. In this context, the chemist needs an approach that will provide a basis of regularity (albeit relying on an essential diversity) in order to be able to tame the promiscuity of chemical nature.

If we consider the example of the element carbon, the distinction made by Mendeleev might seem intuitive or even trivial. In this case, the same element — carbon — can take the form of charcoal, diamond, and

TABLE OF SIMPLE SUBSTANCES.

Simple fubftances belonging to all the kingdoms of nature, which may be confidered as the elements of bodies.

New Names.	*Correfpondent old Names.*
Light - - -	Light.
Caloric - - -	{ Heat. Principle or element of heat. Fire. Igneous fluid. Matter of fire and of heat. }
Oxygen - - -	{ Dephlogifticated air. Empyreal air. Vital air, or Bafe of vital air. }
Azote - - -	{ Phlogifticated air or gas. Mephitis, or its bafe. }
Hydrogen - - -	{ Inflammable air or gas, or the bafe of inflammable air. }

Oxydable and Acidifiable fimple Subftances not Metallic.

New Names.	*Correfpondent old names.*
Sulphur - - -	}
Phofphorus - - -	} The fame names.
Charcoal - - -	}
Muriatic radical -	}
Fluoric radical - -	} Still unknown.
Boracic radical - -	}

Oxydable and Acidifiable fimple Metallic Bodies.

New Names.		*Correfpondent Old Names.*
Antimony -	⎫	⎧ Antimony.
Arfenic - -	⎪	⎪ Arfenic.
Bifmuth - -	⎪	⎪ Bifmuth.
Cobalt - -	⎪	⎪ Cobalt.
Copper - -	⎪	⎪ Copper.
Gold - -	⎪	⎪ Gold.
Iron - - -	⎬ Regulus of	⎨ Iron.
Lead - - -	⎪	⎪ Lead.
Manganefe - -	⎪	⎪ Manganefe.
Mercury - -	⎪	⎪ Mercury.
Molybdena - -	⎪	⎪ Molybdena.
Nickel - - -	⎪	⎪ Nickel.
Platina - -	⎪	⎪ Platina.
Silver - -	⎪	⎪ Silver.
Tin - -	⎪	⎪ Tin.
Tungftein - -	⎪	⎪ Tungftein.
Zinc - -	⎭	⎩ Zinc.

Salifiable fimple Earthy Subftances.

New Names.	*Correfpondent old Names.*
Lime	{ Chalk, calcareous earth. Quicklime. }
Magnefia	{ Magnefia, bafe of Epfom falt. Calcined or cauftic magnefia. }
Barytes	Barytes, or heavy earth.
Argill	Clay, earth of alum.
Silex	Siliceous or vitrifiable earth.

Figure 9. The list of simple substances drawn up by Lavoisier and published in his *Elements of Chemistry* in 1789. Private collection.

fullerenes, three very different simple bodies but all composed exclusively of pure carbon. We need to go further, however, and examine the theories that lie behind practical examples like this. The most important one is Avogadro's hypothesis that forms a bridge between phenomenal (homogeneous or heterogeneous) molecules and their constitutive atoms. From this perspective, Mendeleev's simple body is to the element what the molecule is to the atom. It is this distinction that informs the current schoolbook definition of an elementary substance: it is a collection of molecules formed by atoms of the same species. In similar fashion, a compound is an ensemble of molecules formed out of atoms of different species.

What is particularly noteworthy is that Mendeleev opted for the notion of element above that of the atom as the core concept for chemistry. This choice proceeds neither from an anti-atomistic position nor from any desire to revive Aristotle's metaphysical notion of the element. Far from being hostile to or even sceptical with respect to the hypothetic notions of atoms and molecules, Mendeleev enthusiastically endorsed them when he spoke at the first International Congress of Chemistry in 1860. Furthermore, he referred to this vision as the solid rock on which he built up the periodic system. On the other hand, Mendeleev was a staunch positivist who resisted what he saw as metaphysical speculations and notions. Nevertheless, he realized that Lavoisier's redefinition of elements as simple substances could not provide a satisfactory basis for a system of chemical substances, as it could not guarantee the conservation of material individualities through chemical transformations. Beyond the concrete bodies obtained at the end of the process of analysis carried out in a laboratory and characterized by their physical and chemical properties, chemists cannot dispense with an immutable material entity responsible for the observable properties of compound and simple substances. Elements, in contrast to simple and compound substances have no tangible reality, they are abstract entities that cannot be touched or seen. Georges Urbain, a French chemist claimed that Mendeleev's "element" had an "ideological character" in the sense that it only existed as an idea.

This something shared by a simple body and all its combinations has a distinctly mysterious character. Nevertheless, its real existence seems to be indisputable even though it exists only in the mind. It is this something that I will provisionally call an element. According to this definition, the element transcends the immediate surveillance of the senses. This notion, which undeniably has its origin in experiments, nevertheless has — and I insist on this point — an ideological character.[5]

To admit that Mendeleev's elements have no physical existence does not mean that they are metaphysical entities. Despite their abstract nature, elements are material and not ideal entities, and they are characterized by a quantitative property, their atomic weights, accessible through experiment. They belong to what Mendeleev termed "the solid ground of positive science".

Thus, Mendeleev clearly recognized the need for chemists to rely on an abstract entity in addition to their arsenal of visible or invisible physical units such as molecules and atoms. By placing the burden of explanatory power on the abstract notion of elements rather than on atoms and molecules, Mendeleev made a philosophical choice, what we can think of as a kind of wager. Indeed, the reality of the individuality of the different chemical elements is clear from the system of relationships revealed in the periodic law. Without this abstract notion, Mendeleev could never have predicted the existence of new elements before they could be isolated as simple substances. The phenomenological notion of simple substance allows neither predictions nor the determination of general laws governing the irreducible diversity of chemical phenomena. Thus, the periodic table itself serves to defend Mendeleev's postulate of the individuality of elements against those chemists or physicists who wanted to reduce the specificity of each element to a single "originary" proto-element. By assigning a place to each element, the periodic system anchors the unit-element in a whole network of material, chemical relationships.

In sum, the mind's eye view implicit in Mendeleev's concept of the chemical element can only be attained via the phenomenological data of chemical experiment. The abstract notion of the "element" depends on the individual, concrete chemical substances and their interactions. Behind this thinking lies, if not a paradox, at least a tension characteristic of chemistry. Indeed, Bachelard points to this tension right at the beginning of his essay on chemical philosophy.

It seems to us, in fact, that the chemist's thinking oscillates between pluralism on one hand, and a reduction of this same plurality on the other.

Thus, we can observe that chemistry does not hesitate to multiply elementary substances when observing heterogeneous compounds often produced by the hazard of experimentation. This is the initial stage, that of discovery. Then a kind of bad conscience intervenes and the chemist feels the need to apply a principle of coherence, as much to understand the properties of the compounds themselves as to seize the true nature of the elementary substances.[6]

What Bachelard here presents as a form of repentance motivated by the chemist's bad conscience about endlessly multiplying chemical entities can instead be interpreted as the kind of wager described above. It is a wager with particularly long odds, but one that Mendeleev has not been alone in taking; other chemists have since followed him down this particular path. Thus, the move is not, as Bachelard suggests it is, to compensate for or to moderate the "substantial realism" that threatens to bury the philosophical chemist under a multitude of particular concrete phenomena by adopting a Pythagorean, idealized mathematical model capable of establishing the desired minimalist harmony. Instead, the idea is to keep a discourse on these concrete singularities alive at the heart of the modern sciences, and to flesh out their causes by a process of abstraction, but without attempting to attain an ultimate universal explanation. After all, the reality of the world *is* that we can only ever experience particular, concrete events and substances acting in local circumstances. Nevertheless, the application of our intelligence to such individual situations is capable of generating general laws by separating out the "essential" elements bound up in our experience of reality.

Renewing Mendeleev's Wager

Mendeleev's wager did not have to wait long before being challenged. The discovery of radioactive decay at the end of the nineteenth century, followed by the isolation of isotopes at the beginning of the twentieth threw Mendeleev's schema of elements into doubt. These discoveries coincided with and contributed to the emergence of sub-atomic physics and the development of a model of the atom composed of protons, neutrons and, of course, electrons. The term "isotope" was coined by Frederick Soddy in

1913 to solve a chemical puzzle raised by the production of radioelements. The chemists who were exploring radioactivity struggled to incorporate these newly identified "elements" into the periodic system. They faced a dilemma: if they chose to follow Mendeleev's definition of an element to the letter, they would be obliged to revise the periodic table to make space for these new elements, thereby betraying Mendeleev's legacy in a different sense. Indeed, this was the approach favoured by de Fajans who noted that the new substances had different properties and thus deserved to be treated as different elements. By contrast, Friedrich Paneth and his colleague Georg de Hevesy, who demonstrated that radiolead could not be separated from lead by any chemical method, argued that although they had different atomic weights, these elements nevertheless shared the same chemical identity. Mendeleev's system was shaken to its very foundations: not only were chemical properties not dependent on atomic weights but, more importantly, an element could be constituted by more than one kind of atom. In fact, the chemical puzzle was solved, or at least resolved by the invention of the term "isotope" (same place), indicating the choice to save the periodic table. Mendeleev's choice to place the core of chemical identity in elements rather than in atoms was renewed at the cost of a slight modification in the enunciation of the periodic law expressing a dependence on atomic number (nuclear charge, corresponding to the number of electrons in a neutral atom) rather than on atomic weights. Modern chemistry is undeniably "electronic", since the chemical properties of an element depend on the states of the outer electrons of its atoms, that is to say, their energy levels and spin. Similarly, for compounds, the key factor determining their chemical properties is the participation of electrons in bonds. But does the development of twentieth-century atomic physics require banishing the notion of element from chemistry?

From a modern-day perspective, the use of the atomic number instead of the atomic weight to define an element seems all too obvious, as we know that this atomic number Z, which is the number of protons in the nucleus, is approximately proportional to the atomic weight. Indeed, it is the variation in the number of neutrons in the nucleus of atoms with the same atomic number that gives rise to isotopes of the same element. As the number of electrons in a neutral atom is the same as the number of protons in the nucleus (the atomic number), and the chemical properties depend largely on the number of electrons, it is quite "natural" that the periodicity of chemical properties should reflect the atomic number, which is in turn, as we have just pointed out, closely paralleled by the atomic weight. Nevertheless, this solution was far from obvious at the beginning of the twentieth century, because in order to save the periodic table and the periodic system, it looked as though chemists would have to abandon Mendeleev's definition of an element.

Mendeleev's definition put into play a tacit distinction between the chemical order and the physical order, a distinction attacked by Urbain, who, along with Paneth, was one of the chemists behind IUPAC's new definition of the element. Urbain was opposed to the idea that when Rutherford bombarded nitrogen or phosphorous with alpha rays, he generated only a physical phenomenon and so Urbain rejected the approach that would place radioactive isotopes in the same box in the periodic table as the non-radioactive ones. Furthermore, the phenomenon of radioactive decay demonstrated that even the idea of a simple body that lay behind Lavoisier's understanding of an element was no longer valid. Urbain interpreted radioactive decay as a form of analysis, and because "simple bodies" did not survive the bombardment with alpha particles, they could not be considered truly simple. Nevertheless, Mendeleev's concept of the element managed to escape this particular line of attack. Because it was a conceptual notion, an abstraction, as we have explained above, Mendeleev's element was able to resist the attacks of the most powerful instruments of modern atomic physics.

Overall then, despite a number of disagreements that revealed the conventional character of decisions concerning their fundamental concepts, the chemists who baptized the isotope renewed Mendeleev's wager by reaffirming the pertinence of the concept of the element as a distinct chemical entity, albeit now defined in relation to sub-atomic particles constituent of matter. Evidently, the notion of the element had to be amended to take into account the idea that all atoms were made up of the same protons, neutrons and electrons, and that these combinations were capable of changing due to collision or radioactive decay. However, this did not mean the end of the element as a real and relevant chemical entity.

Friedrich Paneth clearly defined the epistemological status of the chemical concept of element in a lecture bearing this title that he gave in 1931.[7] Here, he raised two questions. The first one recalls Aristotle's problem with "true mixts": "In what sense may one assume that the elements persist in compounds?" His answer — based on a historical survey of the notions of elements and atoms — is clear: since chemistry is concerned with the secondary qualities of substances, chemists should assume the persistence of qualities in compounds. He proposed to name what persists "basic substance" (*Grundstoff*) while he reserved the term "simple substance" (*einfacher Stoff*) for the matter exhibiting the phenomenological manifestations associated with this abstract basic substance, thereby renewing Mendeleev's conceptual distinction. Paneth's second question was: "whether or not it is true that chemistry should and will dissolve into physics?" Here, he dealt with the possibility of reducing chemistry to physics, a development that had already been predicted by some physicists.

Who's Afraid of Reductionism?

In 1929, Paul Dirac, one of the creators of quantum mechanics, declared that the underlying physical laws necessary for the mathematical theory of a large part of physics and the whole of chemistry were now completely known. The only difficulty, according to Dirac, was that the exact application of these laws led to equations much too complicated to be soluble.[8] This and other similar pronouncements clearly demonstrated the ambition of physicists to reduce all chemical phenomena to the quantum mechanics of atoms. The argument itself is quite straightforward. In principle, chemistry is entirely circumscribed by the description of electron distribution expressed by the Schrödinger equation. With the principle established, the rest of chemistry is a matter of details. Even though exact solutions for the Schrödinger equation are only known for simple atoms like hydrogen and helium with only one or two electrons, it is reasonable to suppose that with the improvement of mathematical tools, and in particular, the huge multiplication of computing power brought in by the digital age, better and better approximations or solutions for all the other elements and compounds will follow. Thus, chemistry could, in principle at least, be deduced from quantum mechanics.

One might imagine that these propositions proclaiming the reduction of chemistry to quantum mechanics would have provoked a storm of protest from the chemical community keen to defend its disciplinary autonomy, but no such reaction was seen at the time.[9] Thus, Paneth, for instance, calmly replied that while physicists might try to reduce all chemical substances to an ultimate primary matter, chemists "will still be justified in going no further than the reduction of the phenomena to the chemically indestructible substances — the elements — and thus in retaining qualitative differences in its fundamental concepts".[10] Most chemists, it seems, were not particularly bothered by Dirac's principle of reductionism. When they started citing this article in the 1930s, it was not to consider its programmatic opening paragraph but rather to try to apply the quantum-mechanical model it proposed. One might suppose that the chemists failed to react either because they were blind to the implications of this reductionist project, or because they had already accepted the subaltern position of chemistry it entailed. There is, however, an alternative interpretation of their muted reaction, which is that chemists did not see the reduction to quantum mechanics as a credible project. After all, there was good reason to be sceptical. First, the emergence of atomic physics was to a large extent based on chemistry. Not only had Mendeleev's periodic table served as a guide for interpreting atomic structure, as we have discussed in the case of the introduction of the term isotope, but also Bohr's model of the atom, with its positively charged nucleus orbited by negatively charged electrons, was based on chemical phenomena. The picture of orbiting electrons capable of jumping from one orbital to another either by absorbing or emitting photons was drawn up using chemical absorption and emission spectra and following patterns familiar from the chemistry of the elements.

It was not, therefore, chemists who were shocked by Dirac's reductionist project but rather his fellow physicists who complained that the model of the atom proposed in this approach contravened the laws of physics. Indeed, what they disparagingly referred to as the "second quantum theory" appeared less like an attempt at reducing chemistry to physics than an attempt to redefine the latter in conformity with the former. But the physicists were not too concerned either. Dirac's project failed to address the real difficulties in terms of the procedures for calculation and approximation that needed to be overcome before anyone could offer analytic quantum mechanical treatments of any atoms or molecules apart from hydrogen and helium, let alone their interactions.

While research chemists may not have been intimidated by Dirac's proclamations, it seems that chemistry teachers were, and still are. The critics from this community regarded the ambitions of the physicists as threatening, and consequently sought to reorient the science towards the macroscopic phenomena that had traditionally been at its base. They were wary of what they saw as a hegemonic philosophical project. Indeed, the claim resonated with the long-standing philosophical ambition for the unification of the sciences premised on the discovery of the ultimate explanations for all natural phenomena. Most interpretations of this project, which was a central item in the agenda of the logical positivists, cast physics in the role of the fundamental science, with chemistry being just one of its derivatives.[11]

This fear of the reduction of chemistry to physics has recently re-surfaced in the form of a revival of the philosophy of chemistry in the United States. In 1999, the chemist Eric Scerri founded a new journal, *Foundations of Chemistry,* addressing precisely these kinds of problems. Indeed, the title echoed that of the famous series of publications by the logical positivists in the US, the *Foundations of the Unity of Science*, which, symptomatically, had nothing to say about chemistry. Eric Scerri himself sees the philosophy of chemistry as the major weapon in the fight against the reduction of chemistry to quantum mechanics, the goal being to demonstrate that the foundations of chemistry are to be found in chemistry itself and not in physics.

Paradoxically, the issue of the reduction to quantum mechanics no longer appears to be such a pressing concern. The project of the unification of the sciences by reducing one body of disciplinary theory to another has run up against innumerable obstacles, and has been attacked by at least a generation of philosophers.[12] Those who want to discuss the relationship between quantum mechanics and chemistry do so in much more subtle ways than Dirac. As Monique Lévy has shown in her study of the issue, this case of reduction does not fulfil all the criteria enumerated by Ernst Nagel. Thus, while there may be an asymmetric relationship between chemistry and physics, it is impossible either to deduce chemistry from quantum mechanics or to make testable predictions of most chemical phenomena based on quantum physics. Indeed, quantum mechanics only potentially addresses a small part of what is included in chemistry.

> If we limit chemistry to what could be deduced from physics alone, whole areas of this science would disappear (kinetics, organic chemistry, bio-chemistry, non-equilibrium processes…).[13]

The concepts of electrons, protons, orbitals, and energy levels that are deployed to explain chemical phenomena only succeed in doing so thanks to the judicious use of complementary hypotheses. According to quantum theory, carbon should be bivalent, but this does not stop the major part of its combinations being tetravalent. The hypothesis of the hybridization of orbitals explains this tetravalence, but, while compatible with quantum theory, it remains *ad hoc*. Furthermore, while the systematization of theories in physics might well have succeeded in integrating certain elements of chemistry, it has not put an end to the development of independent chemical theories. That is why Monique Lévy argues in the end for a "reduction by synthesis" which is a form of reduction that reserves a fundamental role for the "reduced" discipline.

If we look at Eric Scerri's approach to the question of reductionism, we see that he not only rejects "hard" reductionism, which aims to explain all chemical phenomena based on quantum mechanics, but also the fall-back "soft" position that posits an ontological dependence without demanding reduction at the level of explanation.[14] Instead, he proposes a third option that consists in defending the autonomy of chemistry by proclaiming that the debate involves different levels of reality that are "autonomous, but interconnected". In particular, he emphasizes the difference between the orbitals as they are used by chemists and those dealt with by physicists working in quantum theory. Indeed, Scerri maintains that chemistry teachers should specify that the orbitals that they are talking about are not those of modern quantum mechanics. Nevertheless, it is probable that this "double language" approach does more harm than good, as it implicitly reaffirms the prejudice that chemistry is a science lagging behind physics. This approach echoes the interpretation of the chemical revolution in the eighteenth century as being chemistry's belated equivalent of the revolution in astronomy and mechanics that had occurred a century earlier. According to this thesis, the "Scientific Revolution" of the seventeenth century founded modern physics, while chemistry had to wait until Lavoisier's reforms at the close of the eighteenth century for the equivalent modernization of chemistry; the so-called "postponed revolution".[15] The modern equivalent of this misreading of history would be for observers to think that modern chemists are still stuck at the first quantum theory, articulating Bohr's simple orbital model, while physicists have long ago abandoned this terrain for better, more advanced models.

To help chemistry develop its own philosophy, chemists need to move beyond this fear of reductionism and focus on other problematics. Quantum theory clearly offers considerable advantages for the teaching of chemistry and, in particular, for presenting the periodic table, with its explanation of how and why the diverse chemical properties of the elements obey an overall pattern. Nevertheless, this does not mean that chemists can get by with quantum mechanics alone. In order to know the chemical properties of a substance, one needs to manipulate it and test it in the laboratory: Does it conduct electricity? What oxides does it form? Is the sulphate soluble in water? Those who claim that all these properties can be deduced from a knowledge of the electron structure of the atoms involved, normally start by qualifying this assertion with an "in principle". This is because when chemists study the properties of elements and other substances in detail, there are always surprises. Furthermore, the exploration of such unexpected properties has led to some of the most interesting chemistry of the twentieth century.

Between the Macro and the Micro

Bachelard once claimed that chemistry was becoming less a science of facts and more a science of effects,[16] but he did not take into account the full potential of the instrumental technologies that had allowed chemists to penetrate the microscopic structure of bodies. The technique of X-ray diffraction as developed by William Bragg in the 1910s opened a window onto the hidden structural world of atoms and molecules. Using this technique, it was now possible if not to "see" at least to deduce the microstructure of metals, to establish relationships between certain measurable macro-properties of matter and the microscopic configuration of the substance under examination.[17]

Once again, however, this technological innovation giving access to the microscopic constitution of matter did not spell the end of the element or the subjugation of chemistry. In 1939, Georges Champetier suggested that the periodic table had effectively been reinvented due to the atomic theory. In his words, "the atomic theory has now become the most solid foundation of theoretical and experimental chemistry. [...] The periodic classification has taken on a much greater significance following the clarification of our ideas concerning atomic structure."[18] Champetier suggested that the reform could be pushed further by redefining the element in terms of the combined number of protons and neutrons in the nucleus, giving every "isotope" its own place in a new table. Despite this proposed re-orientation of the periodic table around discoveries in atomic physics, Champetier never suggested that it might be possible to deduce chemistry from physics, seeing the future relationship between the two disciplines in terms of collaboration rather than domination.

> I have tried to emphasize the evolution of chemistry triggered by the introduction of the physico-chemical disciplines and methods. I wanted to show that the chemist cannot isolate himself from these developments and be satisfied with just carrying out and describing reactions. He will not be able to capitalize on his previous successes unless he assimilates the disciplines of the physicist or agrees to collaborate with him. This evolution of chemistry towards physico-chemistry is the most characteristic feature of the last twenty years.[19]

The relationship between microstructure and macroscopic properties that forms the central axis of solid-state physics has profoundly modified the approach of all the chemists working on solid materials. Today, they dispose of a whole range of instruments to indicate the organization of macroscopic bodies at the atomic level: neutron diffraction; infra-red spectrometry; synchrotrons; electron microscopes; nuclear magnetic resonance; not to mention the scanning tunnelling microscope. Thus, the chemistry of solids continues to become increasingly oriented towards materials science, which studies — or, increasingly, conceives — materials in terms of their properties, developing materials that are particularly well suited to given applications. From this perspective, the chemist can see the periodic table as a well organized toolbox. Each column is like a drawer in this toolbox, with each element providing a certain network of properties. While one group provides a range of ionic metals that are good conductors, they can be used along with the oxides of another group to make a battery. Other groups provide semi-conductors, and the halogens, for example, supply a range of additives to make fluorescent lamps of various colours. The periodic table can be seen as an accessibly organized warehouse, or a catalogue from which one can order the basic elements to construct new materials that fulfil the desired functions. This is certainly the impression given by the chemist Michel Pouchard:

> The chemist is primarily the architect of matter as well as its mason; his scale is that of the nanometer, his bricks the hundred or so elements in Mendeleev's periodic system, and his cement is their valency electrons.[20]

Thus, the notion of the element finds itself vindicated while it is at the same time redefined as providing the "bricks" for a technological project.

This reconfiguration of the element in the twentieth century has given rise to a paradoxical situation. All the members of the chemical community appear to accept the notion — albeit tacit or implicit — of an element when they refer to the periodic table, placing it at the heart of modern chemistry. Nevertheless, the concepts they mobilize in their everyday use of the elements are at best unclear if not outright contradictory. Thus, Mendeleev's abstract notion of an element can, it seems, be put to one side when it comes to getting things done, when the chemist treats the elements as agents or instruments. The irony is that the conceptual purity of the element is preserved in the context of chemical education. It is the students who are being introduced to the discipline who are taught the strictly abstract sense of an element, while a more pragmatic, anything-goes interpretation is the privilege of the experienced chemist.

To sum up, the different uses of the concept of an element impose different exigencies. If the concept is treated as a foundational notion for chemistry, then it must be approached with philosophical rigour, while if it is regarded as a working tool for the practicing scientist, then it can be used as loosely as is consistent with getting the job done. This dichotomy is reflected in the tensions within the chemistry community concerning not only the concept of the element, but also the interpretation of the periodic table itself. Despite these tensions, however, the concept of the element occupies a particularly important position in the pantheon of chemistry, as Mendeleev himself suggested in his own immodest assessment of the value of his work:

> Kant believed that there were "two objects in the universe that provoke the admiration and veneration of man: the moral law within us, and the stellar heavens above us". After further exploring the nature of the elements, we need to add a third such object "the nature of the elemental individuals that is expressed all around us" in light of the fact that without these individuals, we would be unable even to form an idea of the stellar heavens, and that furthermore the notion of these atoms reveals the singularity of their individuality, the infinite repetition of these individuals, and their subordination to the harmonious order of nature.[21]

There is, however, a fundamental difference between Kant's moral law, Newton's laws and the periodic law. Only the last one places no constraints on how it should be interpreted, leaving the periodic table open to multiple readings. It might well be the polysemic nature of the table that lends it its extraordinary vitality. Indeed, none of the many interpretations that have developed over time and in light of subsequent discoveries has succeeded in chasing out Mendeleev's original highly abstract notion of element used in elaborating his table.

The previous reflection brings us back to the stereotype of the opposition between chemistry and physics, with the first concrete and pragmatic, the second abstract and idealist.[22] Nevertheless, we want to emphasize the point that chemistry cannot develop general laws of its own without the process of abstraction. The chemical element remains, therefore, a "material abstraction" that cannot be reduced to the concepts that surround it. There are two senses to this abstraction. First, the element is abstract because it is the result of an active effort to detach significant relevant features from the particular local circumstances in which the elements exhibit their chemical action. No two reactions are identical in every detail, but the salient features of a series of similar reactions can be abstracted by the senses of experienced chemists with or without the aid of their instruments. But there is a second and less concrete form of abstraction involved in the construction of the explanatory structure of the periodic system. The idea that macroscopic chemical properties depend on an invisible causal factor (sub-atomic structure whether it is indicated by atomic number or atomic weight) that can be inferred based on a theoretical construction represents another significant form of abstraction. In this respect, the chemical element is close to a mathematical abstraction. Thus, it becomes an instrument that allows one to construct a series that is assumed to lie behind and to be more fundamental than all individual, local, observable chemical changes. This abstraction allows the chemist to describe the order behind the apparently chaotic multiplicity of the phenomenal world.

The greatest value of Mendeleev's concept of an element is its power to construct relational series. Chemical elements are effectively "disembodied" abstract entities that are attached to the real world by the network of relationships described by the periodic system. This abstract-concrete relationship is represented by the element's position in the periodic table, which in turn forms the basis for a multiplicity of interpretations.

References

1. See B. Bensaude-Vincent (1986).
2. The theory-laden nature of empirical observations is an argument that philosophers have mounted against a naïve view of experiment as providing a supply of objective, unbiased facts with which to test a theory. In general it fits into a more global position associated with the name of Pierre Duhem that it is impossible to falsify or confirm one single statement in a physical theory. This position, which was reformulated by Willard van Orman Quine, is known as the Duhem-Quine thesis and argues that all the parts of modern physical theory are interconnected, and it is therefore impossible to test any constituent hypothesis in isolation.
3. A.-L. Lavoisier (1789), transl 1790, p. xxiv.
4. D. Mendeleev (1871), p. 693. Mendeleev's distinction between a simple body and the atom represents an important reorientation in chemistry. At the end of the eighteenth century, the discipline was centred on analysis, but it would subsequently become reoriented around the relationship between the observable,

measurable properties of bodies and the determinant cause of these properties. The goal in this context was the articulation of the phenomenological behaviour of each one of a range of bodies (the phenomenal elements) with an abstract 'thing' identified by its atomic weight but not to be confused with the atom in its traditional philosophical sense.

5. G. Urbain (1925), p. 9.

6. G. Bachelard (1930), p. 5.

7. F. Paneth (1931). See also K. Ruthenberg (1993).

8. P. A. M. Dirac (1929), p. 714.

9. A. Simoes (2002). See also, K. Gavroglu and A. Simoes (1994).

10. F. Paneth (1931), p. 160.

11. E. Nagel (1961) and (1970).

12. P. Feyerabend (1965), P. Galison and D. J. Stump, eds (1996).

13. M. Lévy (1979), p. 348.

14. E. R. Scerri (2000).

15. This phrase was coined by Herbert Butterfield in H. Butterfield (1957).

16. G. Bachelard (1930), p. 229.

17. For more on the history of solid-state physics, see L. Hoddeson, *et al.* eds (1992).

18. G. Champetier (1940), pp. 9–10.

19. Ibid, p. 30.

20. M. Pouchard (2003), p.6.

21. D. Mendeleev (1889), p. 987.

22. See N. Cartwright (1983) for more on the abstraction inherent in the approach adopted by physics.

CHAPTER 10

POSITIVISM AND CHEMISTRY

One of the great philosophical questions that has challenged mankind at least since the first written traces of philosophical inquiry is: What can one know?

There are two parts to this question, one ontological, the other epistemological. Thus, the ontological question is what exactly is there to know? What makes up the world? The epistemological question is what out of the range of things that make up the world can we know? Of course, these two questions are intimately associated, as philosophers are constrained in their ontological thinking about what makes up the world by their mode of knowing, be it divine revelation or scientific empiricism.

Since Kant, it has become more or less standard for philosophers to base their response to the most fundamental epistemological questions on physics, and more recently, the cognitive sciences. Physics was seen as presenting the "royal road" to ultimate knowledge, and as a corollary, would define the limits of our capacity to know. Perhaps the clearest message delivered by mechanical philosophy was that there was an ultimate explanation for the functioning of the world. It was assumed that the ultimate particles that composed the springs and levers were the "real" explanation behind all observable phenomena. In Fontenelle's terms, there was an "end" to scientific analyses.[1] This is the conceptual context in which the classic debate between realism and positivism is generally posed by philosophers of science. Is there an ultimate underlying mechanism responsible for the phenomenal world or not?

This way of thinking about the problem of realism corresponds to what Isabelle Stengers has termed "the faith of the physicist".[2] This credo, championed by Max Planck and Albert Einstein among others, expresses a faith in an accessible intelligible world, a cosmic order independent of human interests and practices. This belief, however, is not helpful for thinking

175

about the philosophy of chemistry we are proposing here. We need, therefore, to reconsider this question of realism from another perspective.

A Variety of Positivisms

With its refusal to engage in the search for the ultimate explanation and its unwillingness to pronounce on the "real" causes that lie behind the world of appearances, chemistry might be considered the science that best exemplifies the principles of positivism.

To understand this claim, we need an understanding of what is meant by positivism. In general, when today's philosophers talk of positivism they think of a hard-headed empiricism that refuses to go beyond the evidence of the senses. Historically, the term "positivism" was coined by Auguste Comte in 1848 and was derived from Comte's description of "positive science" as the following: real (as opposed to dealing in fanciful or abstract metaphysical entities); useful (as opposed to pointless speculation); precise (as opposed to vague); positive (as opposed to negative), and finally relative (as opposed to absolute).[3] Although Comte advocated the use of hypotheses and heavily criticized Bacon's empiricism, the term positivism gradually came to be used as a synonym for Baconianism. It was only towards the end of the nineteenth century, when it was applied to the philosophy of science developed by Ernst Mach, that the term "positivism" came to be used as an antonym of realism. Mach's positivism consisted in rejecting the Platonic vision of a "real" world lying behind and underwriting the phenomenal world and asserting instead that there is only one single world of empirical sense data. From this perspective, the function of scientific theory is to provide a functional and economical symbolic description of the phenomena and not to describe the ultimate reality that lies behind or beyond the empirical phenomena.

To what extent can chemistry be considered a positivist science? To be sure, chemists have often claimed their attachment to the solid bedrock of empirical phenomena. Indeed, the nineteenth-century French chemist, Jean-Baptiste Dumas, saw this empirical orientation as the timeless essence of chemistry:

Modern and ancient chemists have one thing in common; their method. What is this method, as old as the science itself, and which has characterized it from its very beginnings? It is a total faith in the testimony of the senses; it is an unbounded confidence placed in experience; it is a blind submission to the power of facts. Ancient or modern, chemists want to see with the eyes of their physical body before employing those of the mind: they want to make theories for the facts and not seek out facts for any preconceived theories.[4]

There are several questions raised by this suggestion of a special affinity between chemistry and positivism. First, is it true that chemistry just sticks to the facts? Second, does this attitude really qualify as positivism? These questions are important for understanding the philosophical status of chemistry, as historians of chemistry have repeatedly argued that the overarching influence of positivism stopped chemists from developing their own ontology, thereby rendering the science unworthy of serious philosophical attention. To respond to these questions, we need to have a clearer picture of positivism than the very general principles outlined above. To help clarify this position, we will consider the work of the initiator of this philosophical trend — Auguste Comte.

Chemistry as the Model for Positive Science

Comte's philosophy is based on a bold narrative of the evolution of human thought. For Comte, in the course of its development, the human spirit naturally progresses from the theological (or fictive) state via the metaphysical (or abstract) state to the scientific (or positive) state. This is both a large-scale history describing the development of the way of thinking about the world from the dawn of civilisation to the dawn of the positive era, and a description of a natural progression that can be observed in the thinking of every human being. At the beginning of his *Course on Positive Philosophy*, Comte offers the following account of the three states:

In the theological state, the human mind essentially orients its research towards the innermost nature of beings, the first and final causes of all the effects that occur, in a word, towards absolute knowledge, representing phenomena as products of the direct and continuous action of supernatural agents [...].

In the metaphysical state [...], the supernatural agents are replaced by abstract forces [...] conceived as being able to cause all the observed phenomena by themselves, leading to explanations that consist in assigning a corresponding entity to each phenomenon.

Finally, in the positive state, the human mind recognizes the impossibility of obtaining absolute notions, renounces the search for the origins and destiny of the universe, as well as for the inner causes of phenomena, and seeks exclusively to discover, by means of the combined use of reason and observation, the effective laws of these phenomena, that is to say their invariable relationships of succession and similitude. Thus, the explanation of facts, reduced to its real terms, consists only of the relationship established between diverse individual phenomena and a few general facts, which the progress of science tends to reduce in number.[5]

Chemistry appears to be the model science for this positive state described by Comte. Chemists more than any other scientists have railed against the evils of theoretical speculation as well as any pretensions to penetrate to the innermost nature of things or to seek out the first causes of phenomena. Indeed, this sceptical strand is so marked in the history of chemistry that it is tempting to think that chemistry invented the position of positivism even before Comte had given it a name. The chemical notion of "salt" offers an exemplary illustration of the passage of a chemical concept from the metaphysical to the positive era. In the seventeenth century, salt was a singular term that designated the universal principle responsible for the materiality of natural things. In the eighteenth century, the term "salts" started to be used in the plural as, in light of empirical evidence they came to be understood as compounds made up of an acid combined with an alkaline part.[6] Thus, once one comes to consider the formation of bodies in terms of relationships rather than just the essential nature of the principles involved, the conceptual nature of these principles itself is displaced from the field of the substantial to that of the relational. A clearer illustration of Comte's positive ideal is provided by Lavoisier's definition of an element, which we have already encountered in another context in Chapter 9. This new definition not only rejects the quest for the first principles of nature in order to facilitate the progress of chemistry, but also explicitly condemns previous approaches to the problem of determining the elements as metaphysical.

> All that can be said upon the number and nature of elements is, in my opinion, confined to discussions entirely of a metaphysical nature. The subject only furnishes us with indefinite problems, which may be solved in a thousand different ways, not one of which, in all probability, is consistent with nature. I shall therefore only add upon this subject, that if, by the term elements, we mean to express those simple and indivisible atoms of which matter is composed, it is extremely probable we know nothing at all about them; but, if we apply the term elements, or principles of bodies, to express our idea of the last point which analysis is capable of reaching, we must admit, as elements, all the substances into which we are capable, by any means, to reduce bodies by decomposition.[7]

According to a well-established tradition, Francophile historians of chemistry cite this definition as marking the beginning of modern chemistry, whereas Anglophile historians see the founding act in a similar definition given by Boyle in his *Sceptical Chemist* from the seventeenth century. No matter who is credited with the foundational act, what is important is the inauguration of a positive, quantitative, experimental science, freed from the dead weight of alchemical superstition. Nevertheless, looking at Comte's writings, in particular, his *Course of Positive Philosophy* published less than fifty years later, chemistry is not given pride of place, with Comte crediting other sciences, like astronomy and even biology as his principal sources of inspiration. Comte criticizes chemistry for being largely a descriptive and not a predictive science. Indeed, he argues that chemistry scarcely merits the title of science at all, as it is a simple collection of facts.[8] Interestingly, the only redeeming feature of the history of chemistry in Comte's view was the new nomenclature introduced by Lavoisier, Guyton de Morveau, Fourcroy and Berthollet in 1787, but even this failed to raise it to the rank of the other experimental sciences.

Positivism as an Obstacle?

Comte's own lack of respect for chemistry has not, however, proven any obstacle to the forging of a strong historical link between this science and positivism. A number of historians have argued that the resistance of French chemists to the atomic theory during the nineteenth century was due to the retrograde, conservative influence of positivism in France.[9] This thesis of the negative influence of Comte's positivism in France has also been used to explain Dumas's notorious statements about atomism in his lesson on the subject:

> Were I the master, I would efface the word *atom* from the science, as I am convinced that it goes beyond experience and we should never go beyond experience in chemistry.[10]

While it is certainly true that many French chemists were reluctant to accept the atomic theory, the explanation that this was due to the influence of Comte, despite its superficial plausibility, remains unconvincing on several accounts. As one of us has argued on several occasions, there are a number of problems with the more indirect spirit of the age, or "*Zeitgeist*" argument.[11] First, there is the question of why such an influence would affect only chemists, and not physicists or other scientists. More significantly, this argument depends on a double misunderstanding; first, a misinterpretation of the tradition of French positivism, and second, a misrepresentation of the construction of modern chemistry across the nineteenth century.

First, let us consider Comte's positivism. The question that interests us here is the following: Was this philosophical position, as it was elaborated by Auguste Comte during the 1830s and 1840s, of such a nature as to lead chemists to deny the real existence of atoms?

In introducing this subject, we have already suggested certain philosophical conclusions concerning elements and realism that are not warranted by the evidence. Lavoisier's famous definition cited above, while it puts into play what might be termed a positivist scepticism concerning the accessibility of the ultimate components of the material world is not necessarily anti-realist. This is the case in two different ways. First, it is not because an "element" is pragmatically defined as a provisional last unit of analysis that it is any less real. Lavoisier's caloric, for example, while it can only be measured and not directly observed (heat is an effect of caloric which is always associated with another substance, rather than the caloric itself) is very real for Lavoisier. Second, it is not the case that the ultimate constituents of substances are not real just because they are not yet known. Furthermore, Comte himself appreciated the chemists' atomism as he knew it in the 1830s, and far from condemning John Dalton's atomic hypothesis, he had nothing but praise for this modern genius.[12] He saw the hypothesis not only as an extension of the doctrine of definite proportions, but also as a chemical version of the corpuscular theory, a doctrine that Comte regarded as practically self-evident if not axiomatic. For him, corpuscular theory underwrote the Newtonian synthesis, putting its validity beyond question. Indeed, the programme that Comte proposed for chemistry was to treat the issue of the "real manner in which the elementary particles agglomerate". Thus, Comte encouraged the use of "hypothetical" particles for this combinatorial project, arguing that "the degree of indeterminacy concerning the intimate constitution of bodies" that is necessarily implied by the nature of this research requires chemists to use this liberty in order to represent the arrangement of chemical combinations in the form of binary compounds, if that can facilitate their work.[13] While this idea of indeterminacy and the liberty of theorizing brings Comte closer to later anti-realist incarnations of positivism, which we will discuss more below, Comte was himself never anti-atomist, nor fundamentally anti-realist.[14]

Positivism versus Realism

What then, we might ask, is the origin of this idea that the rejection of atomism by French chemists in the nineteenth century was due to the influence of positivism in this country? In fact, this idea comes from an anachronistic projection of the second notion of positivism that emerged at the end of the nineteenth century and is associated with the names of Ernst Mach, Pierre Duhem, Henri Poincaré, Gaston Milhaud and Edouard Leroy, among other thinkers.[15] This form of positivism was not at all interested in providing a blueprint for human development but rather took on quite a specific philosophical question: the power of scientific theories to attain the real. Their alternative vision of scientific theories was as systems of symbols intended to organize or rationalize our phenomenal experience but without transcending it.

This new positivist movement arose from a retrospective criticism of rational mechanics, based on the application of non-Euclidean geometry to modern physics, as well as being largely inspired by the rise of thermodynamics. Due to the importance of thermodynamics among its considerations, this new positivism engaged directly with a controversy in the late nineteenth-century that pitted atomists against energetists. The dispute has its origins in Wilhelm Ostwald's ambitious project to reinterpret the whole of chemistry in terms of energy, relying on two principles of thermodynamics — the conservation of energy (the first law of thermodynamics), and Carnot's principle that any useful work implies the dissipation of energy (the second law of thermodynamics). Ostwald believed that this approach would allow him to dispense with the atomic hypothesis, and led him to claim that the concept of matter as proposed by the mechanical philosophy was no longer necessary.[16] This programme was harshly criticized in Germany, notably by Max Planck, Felix Klein, Walter Nernst, and Ludwig Boltzmann who maintained the necessity of atomism. Ostwald's 1895 paper translated into English as "The conquest of scientific materialism" was published in French under the title *La déroute de l'atomisme* (*The Rout of Atomism*). This paper prompted vigorous responses not only from atomist physicists like Marcel Brillouin, Paul Langevin and, above all, Jean Perrin, but also from the philosopher Abel Rey.[17] According to Rey, only realist, atomist mechanics that sought to explicate the ultimate nature of phenomena could offer a truly creative science. He objected to this energetist approach that refused to seek out any such ultimate explanations, preferring mathematical formulae over concrete representations.[18] Thus, Ostwald's project to reduce chemistry and physics to the principles of thermodynamics provoked a conflict between two styles of science, one explanatory, causal and realist, the other descriptive, conventional and anti-realist. These were the conflicts that shaped the philosophical landscape in which Emile Meyerson elaborated his own thought, in particular, as it was presented in his *Identity and Reality* from 1908.

Other contemporary debates served to entrench this new positivism, in particular the controversy taking place in Germany that opposed Ernst Mach and Max Planck concerning the aims of science. In France, the beginning of the twentieth century witnessed a debate over the supposed bankruptcy of science, in which energetism was branded an anti-science movement. Whether considered a good or a bad thing, energetism was generally perceived as the purest expression of positivism, seeking to limit the ambitions of science to a description of experience or at best its formalisation. Thus, the controversy that pitted energetism against atomism served to establish an opposition between positivism and realism that lies behind the modern debate in the philosophy of science. Indeed, as Ian Hacking has already noted, the very notion of scientific realism and the philosophical debates around it were shaped in this period.

> Yet realism about atoms and molecules was once the central issue for the philosophy of science. Far from being a local problem about one kind of entity, atoms and molecules were the chief candidates for real (or merely fictional) theoretical entities. Many of our present positions on scientific realism were worked out then, in connection with that debate. The very name "scientific realism" came into use at that time.[19]

References

1. Today, with computing metaphors supplanting mechanical ones, this form of determinism is usually presented in terms of a code or program that provides all the instructions for the development and behaviour of animate and inanimate objects. This model is, of course, applicable not only to physics but also to molecular biology. For a criticism of this position in physics, see P. Jensen (2001).
2. I. Stengers (1997), particularly volumes 2 and 3.
3. A. Comte (1844), pp. 120–125.
4. J.-B. Dumas (1837), p. 4.
5. A. Comte (1830), vol. I, pp. 4–5.
6. F. L. Holmes (1989).
7. A.-L. Lavoisier (1789) transl., 1790 p. xxiv.
8. A. Comte (1830-1842) Lesson 35, Vol. 2, p. 7 ff; see also B. Bensaude-Vincent (1994).
9. A. Rocke (1984), pp. 181–182, D. Knight (1967), p. 105 and p. 126, J. Jacques (1987), pp. 195–208, A. Carneiro (1993), p. 85, and M. Scheidecker (1997). In M. Malley (1979), Marjorie Malley describes positivism as an ambient obstacle that prevented Marie Curie from finding an appropriate theoretical interpretation of radioactivity.

10. J.-B. Dumas (1837), p. 246.
11. B. Bensaude-Vincent (1999).
12. A. Comte (1830–1842), lesson 37, vol. 3, pp. 145–147.
13. *Ibid*, lesson 36, p. 79 ff.
14. The similarities between Comte's and later versions of positivism have been emphasized by Gaston Mihaud, see A. Brenner (2003), p. 33.
15. A. Brenner (2003).
16. W. Ostwald (1895), for a survey of the controversy see E. Hiebert (1971).
17. W. Ostwald (1895).
18. A. Rey (1908).
19. I. Hacking (1983), pp. 30–31.

CHAPTER 11

ATOMS AS FICTIONS

The opposition of nineteenth-century French chemists to atomism is difficult for many philosophers and historians to understand except when it is presented as a stubborn refusal to look objectively at the scientific evidence, for ideological or philosophical reasons. Nevertheless, this difficulty in understanding this opposition to atomism arises from another amalgam, this time concerning not positivism but atomism. The atomism proposed by certain organic chemists of the time as the basis of a controversial structural theory was not the atomism of the early-twentieth-century physicists like Rutherford and Bohr. The physicists' atom was the answer to the classic question concerning the structure of matter, and, as such, has focused philosophers' attention to the exclusion of all others. This exclusive focus on the physicists' atom has the advantage of providing the last step in the classic version of the positivism–atomism fable outlined above. In this view, the experimental demonstrations of Jean Perrin, followed by the elucidation of sub-atomic structure by Rutherford and Bohr were finally able to force the reluctant positivist chemists to accept the reality of atoms.

The atom that was the subject for debate among chemists in the nineteenth century was not the atom of twentieth-century physics; whether they admitted it or not, all of these chemists had a corpuscular vision of matter. Dalton's atom was intended to account for the discontinuity in combining proportions observed in chemical reactions, and not to provide a theory concerning the structure of matter. Thus, we need to consider the atomism of the chemists with respect to their own project rather than in terms of the physicists' atom. If we adopt this approach, we see that there is no need to invoke the influence of Comte and his positivism in order to understand that their interpretation of the laws of chemical combination do not require any hypothesis concerning the ultimate indivisibility of matter.[1] As we have already remarked, John Dalton's atomic hypothesis did not concern the minima of matter, but the minimum units of chemical combination. The central concerns of this approach were how to write appropriate formulae for the compounds, and the corresponding disposition of the atoms in the molecule, which depended on determining the atomic weight for each of the elements concerned. Nineteenth-century chemists were quite clear concerning the role of their atomism, and distinguished what interested them from the question of the divisibility of matter. Thus, for example, Jean-Baptiste Dumas dedicated a whole lesson of his course in chemical philosophy to explaining that the issue of the existence of atoms had nothing to do with chemical atomism, which he proceeded to present in the following lesson.[2]

Writing Formulae

The controversies that pitted chemists against one another over atomism were not fundamentally about the existence of atoms or the continuous or discontinuous nature of matter, but were about the distinction between atom and molecule that formed the basis of Avogadro's hypothesis. The two opposing camps, the equivalentists and the atomists, disagreed over the formulae that should be used to express the constitution of compounds, in particular with respect to the proportions of the constitutive elements. What was at stake was the organization of hundreds of organic compounds with the opposing positions giving rise to very different classifications. The equivalentists adopted a system of atomic weights based on combining equivalents (to take water as an example, in terms of weight, one unit of hydrogen combined with eight units of oxygen to make nine units of water, meaning the atomic weights of hydrogen to oxygen were in the proportion of one to eight), while the atomists like Auguste Laurent and Charles Gerhardt developed an alternative system of atomic weights around equivalents of substitution in compounds. Indeed, to be more precise, the two camps disagreed over the interpretation of combination, with Laurent developing a theory of organic combination based on the idea of the substitution of "equivalents" in a "fundamental radical" or "nucleus" represented by a nine-sided prism.[3] Laurent did not, however, believe that this concept of nucleus, borrowed from Haüy's crystallography, corresponded to any physical reality. Nor was the nucleus a particular material object, as it remained the same even when its constituent atoms had been substituted for others, and it did not even represent any specific fixed form. As Marika Blondel rightly pointed out, the nucleus was that on which one could perform operations; additions and substitutions: "In concrete terms, the nucleus was really nothing except the basis for performing operations [...]. It was, in a certain sense, an algebraic object."[4]

Types and Models

What is most striking in this controversy is that the champions of the atomic hypothesis adopted essentially positivist epistemological positions, and refused to offer any ontological interpretation of their structural formulae. This was true, for example, of Gerhardt who developed a type theory in which all mineral and organic compounds derived from four fundamental types: the hydrogen type H-H; the hydrochloric acid type H-Cl; the water type H-H-O; and the ammonia type H-H-H-N. Gerhardt was able to interpret many reactions using these four types, in particular reactions in which radicals were exchanged by a process of double decomposition. These types allowed him to class organic compounds into four groups, and led to predictions of hitherto unknown compounds formed by the substitution of various radicals for the hydrogen in the different types. Gerhardt conscientiously avoided any realist representation of the internal architecture of these compounds and refused to think of radicals as bodies that could be isolated in any permanent and stable form. For him, a radical was simply "the rule according to which certain elements or groups of elements are substituted for one another or are transported from one body to another in the event of a double decomposition". Gerhardt was careful to point out that his radical had no material reality, and was just a taxonomic schema, suitable for revealing analogies and homologies between chemical compounds. To underline the fact that his formulae had no ontological import, he named them "rational formulae" and even accepted the fact that the same substance could have several different formulae.

The role of a positivist approach in this context is interesting, and it certainly does not seem to have represented an obstacle to chemical atomism. Instead, this positivist approach seems to have given the chemists a considerable degree of liberty in dealing with these issues, thus allowing them to put together all sorts of theoretical or practical arrangements without having to contend with the consequences of these constructs being real. When Wilhelm August Hofmann introduced molecular models made of sticks and balls to visualize the spatial arrangement of carbon compounds, he did so in this spirit of ontological liberty, although this did not stop other chemists, like Benjamin Brodie from complaining about this materialization of theoretical entities. In general, however, chemists continued to use such models — both physical and pictorial — with a tacit understanding that they were purely symbolic rather than representing any substantial reality. These models were treated as tools that could be manipulated, a way of reflecting on the relationship between the elements that composed a molecule.

Kekulé was particularly partial to his story of the discovery of the structure of benzene in a dream. Thus, he recounts how, falling asleep over his chemistry texts, he saw the atoms start to jump and dance in front of him. The atoms of the benzene molecule undulated like a snake, and then the snake-like string of atoms turned around to bite its own tail, thereby forming a closed structure, like the ouroborus of alchemical tradition. Twenty years later, on the occasion of a dinner organized by the German Chemical Society to celebrate the discovery of the hexagonal form of benzene with its alternate double and single bonds, the organizers published a famous pair of drawing in which the six carbon atoms of benzene were represented by a circle of six monkeys. In one version, each of the monkeys holds the next one by the feet, with alternate monkeys using one hand to hold one foot (single bond) and both hands to hold both feet (double bond). In the second illustration, each monkey holds hands with its two neighbours and then makes the double bond on only one side using its tail (Figure 10). This example illustrates quite clearly that chemists did not lack imagination concerning how to visualize their atoms. In his *Lehrbuch der organischen Chemie* (*Textbook of Organic Chemistry*), Kekulé represented the carbon atoms using forms that he aptly termed "sausages" with the length of each one corresponding to its valence. This mode of illustration was not intended to represent the building blocks of chemical compounds, but rather to help understand their mode of bonding, which explains the emphasis placed on the representation of valence. For Kekulé, atoms were essentially the units capable of entering into bonding relationships with one

Fig. 1. Fig. 2.

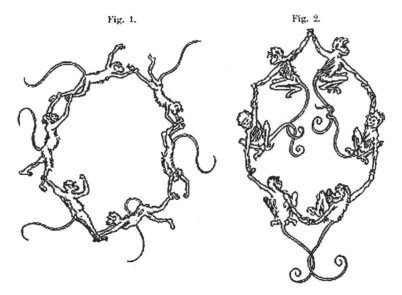

Figure 10. Representation of the benzene ring structure using monkeys, with both figures illustrating the alternating double and single bonds proposed by Kekulé. From the *Berichte der durstigen Chemischen Gesellschaft* (*Annals of the Thirsty Chemical Society*) a special commemorative edition of the *Berichte der Deutschen Chemischen Gesellschaft* (*Annals of the German Chemical Society*), 1886. Private collection.

another, and so his atomism was tied to a theory of atomicity, with each atom characterized by its valence above any other property.

Agnostic Atomism

As for the question of the actual existence of atoms, Kekulé refused to pronounce on the subject and followed Lavoisier in dismissing the question of the nature and number of the ultimate elements of matter. His opinion was that this issue was not a legitimate concern of chemistry and should be banished to the realm of metaphysics. When he was prepared to venture onto this metaphysical terrain, however, Kekulé was inclined to believe that atoms did not exist.

The question whether atoms exist or not has little significance from a chemical point of view: its discussion belongs to metaphysics. In chemistry, we have to decide whether the assumption of atoms is a hypothesis adapted to the explanation of chemical phenomena. More especially, we have to consider the question whether a further development of the atomic hypothesis promises to advance our knowledge of the mechanism of chemical phenomena. I have no hesitation in saying that, from a philosophical point of view, I do not believe in the actual existence of atoms, taking the word in its literal signification of indivisible particles of matter. I rather expect that we shall some day find, for what we now call atoms, a mathematico-physical explanation, which will render an account of atomic weights, of atomicity, and of numerous properties of so called atoms. As a chemist, however, I regard the assumption of atoms, to be not only advisable, but absolutely necessary in chemistry.[5]

Kekulé was not the only chemist to adopt such a position, and this refusal to accept the existence of atoms continued even after physicists and chemists had provided convincing evidence for the reality of molecules. Writing in the twentieth century, the French chemist Georges Urbain continued to treat the atom as a symbol, a simple representation. In general, he refused to address the ontological question of the real existence of atoms on the grounds that it was beyond his competence as a chemist.

The current atomic theory, like all good physical theories, provides an economical way of thinking and relieves our memory. The theory is useful because these images provide a synthesis of a considerable number of relationships that exist between sensible qualities. It is useful because the language that brings these images to mind is clear and adapted to the facts with an adequate degree of precision. These images are like a form of writing composed of symbols that evokes the laws obeyed by the facts. Philosophers can discuss the question of whether atoms have a reality external to us. The study of objects in themselves is not part of science. Science only establishes the relationships that exist between the sensible properties of the delimited portions of space that we call bodies.[6]

In the end, there were simply too many chemists who refused to adopt a position concerning the reality of atoms for us to be satisfied by as nebulous an explanation for this phenomenon as the "intellectual environment" or the *Zeitgeist* in France. This is even less satisfactory as an argument if we believe that not adopting a full-blooded realist atomism presented an obstacle to scientific progress during this period. Thus, we need to look deeper into the reasoning behind this generalized agnosticism concerning the reality of any putative ultimate elements or first principles of matter. What status can we assign to these objects that chemists acknowledged to be fundamental to their science while refusing to enter into the debate over whether or not they existed?

In order to understand what was at stake in this refusal to address the ontological question of atomism, let us consider Kekulé's case more closely. How could Kekulé doubt the existence of atoms and molecules while at the same time conceiving the molecular architecture of benzene that not only enabled him to explain the properties of aromatic compounds, but also allowed his students to synthesize a whole range of new compounds? The situation appears so paradoxical that it is tempting to dismiss the chemist's declarations concerning the existence of atoms as so much cautious rhetoric inspired by the reigning climate of positivism. This was precisely the interpretation given to Kekulé's proclamations by Emile Meyerson, an organic chemist who trained with Bunsen before pursuing a career in the philosophy of science. "He (Kekulé) sometimes voiced his reservations, but only, one gets the impression, reluctantly, for the form. In his heart, he believed firmly in the existence of atoms, molecules and bonds, and he manipulated them exactly as if they were objects that could be experienced with the unaided senses."[7] Meyerson's indignant refusal to take Kekulé's sceptical declarations seriously should help us to recognize the value of the position adopted by chemists like Kekulé; that metaphysical considerations had no influence over the way they did their chemistry.

Atoms as Mediators

To believe or not to believe, that was not, apparently, the question. Doubtless, every chemist believes in the existence of atoms as physical entities, but that is not what is at issue in chemical atomism. The function performed by Gerhardt and Kekulé's atoms was first and foremost logical. These atoms connected together the phenomenal relationships expressed by certain laws — Gay-Lussac's, Dulong and Petit's, Avogadro's — rendering these laws intelligible by providing an overarching schema into which they could fit. From this perspective, Ernst Cassirer, a philosopher of mathematics, seems to have understood the particularity of nineteenth-century chemical atomism better than Meyerson, the ex-chemist. According to Cassirer, the atom is never a pre-existing "thing", but rather the end point of the chemists' claims. Thus, a given property is attached to an atom as though to some absolute "support". Nevertheless, Cassirer suggests that the chemists' aim is not principally to relate the atom to a set of observed properties but rather to draw analogies between such sets of properties. Thus, the atom serves as a mediating concept in building up a global network of phenomenological properties.[8] In other words, the concept of the atom is a tool that brings the multiple relationships between empirical data to a focus, thereby allowing the chemist to unify them. It is not, therefore, a "thing"; it is not an object but rather a subject in the sense of a hypothetical substrate for a host of empirically established relationships. Furthermore, as Cassirer pointed out, this "synopsis" is not only a recapitulation of the facts, but can also predict new ones.

Neither Laurent's nucleus, nor Gerhardt's type formulae represent the organization of atoms in a compound. Instead, they express the capacity of units or modules to enter into relations with others. Thus, the formulae they developed for a given compound indicated the range of possible substitutions. It is clear that in order to think about and represent this capacity to enter into relations with other bodies, the chemist is not obliged to operate with spherical atoms with an arm for each unit of valence, however useful such representations might have been for teaching or amusing chemistry students in the context of the lecture hall. It is interesting to note that the notion of atomicity as defined by Wurtz turned around elements and not atoms, although the capacity to combine with other elements that characterized this property was thought to reside in the atoms, hence the name. Thus, chemical reactions and chemical compounds resulted from the tendency of atoms of a given element to exhaust their capacity to combine with others, a capacity that "resided" to different degrees (depending on the element) in each one of them.[9] The chemist's atom is, therefore, first and foremost a capacity to enter into relationships with others, a substrate of multiple relationships, quite independent of the structure of matter, granular or otherwise. The primary goal of the chemist is to unravel the dense network of relationships between and around bodies, behind which the search for the hidden structure of matter comes a distant second.

The Phenomenalist Response

All nineteenth-century chemists, whether atomists or equivalentists, could agree on the conception of the atom as a symbol representing a range of inter-related phenomena, or a node in a complex network of relationships. If this was the case, why then was there any conflict between these two groups? The disagreement was over related epistemological issues that can be seen in a celebrated debate that pitted the atomist, Charles Adolphe Wurtz, against Marcellin Berthelot, a notoriously influential anti-atomist chemist. The scene for this clash was a gathering of the French scientific elite at the French Academy of Sciences in 1877. Henri Sainte-Claire Deville opened the debate by declaring that Avogadro's law, which distinguishes between atoms and molecules and was used by atomists to help determine atomic weights, was an unfounded hypothesis. Berthelot, champion of the equivalentists, also treated Avogadro's law as a simple hypothesis, and then attempted to ridicule the atomist position: "who has ever seen a gaseous molecule or an atom?"[10] The message behind Berthelot's jibe was that science should be limited to the study of observable phenomena, making him a "positivist" in the modern understanding of the term, albeit with unrealistically high, if not naive epistemological standards for scientific reality. Unsurprisingly, the equivalentist position that Berthelot was defending against the atomists was not devoid of hypotheses itself, as it required a number of implicit hypotheses concerning the nature of chemical combinations in order to operate. Indeed, Wurtz defended himself by pointing out precisely this hypocrisy in Berthelot's position: "At its base, your notation in terms of equivalents covers the same idea of small particles, and you believe in them like we do [...] without repudiating hypotheses entirely, because no science can do without them, no scientist can refrain from making them."[11]

Refusing to accept any hypotheses represents a radical form of positivism, one that cannot, as we have seen, be attributed to Comte himself, but one that goes hand in hand with a certain naïve form of scientific realism (although neither position implies the other). Maurice Delacre, a former student of Wurtz, continued to defend this position in his 1923 publication, *An Essay on Philosophical Chemistry*. Delacre had enjoyed a successful research career in organic chemistry, working in particular on the synthesis of benzene, but had nevertheless reached sceptical conclusions concerning atomism in particular and theory in general. Thus, he contended that "chemistry is a positive science, its positivism is experimental, and it alone, out of all the sciences, possesses such a high degree of positivism. Chemistry can allow itself to refuse all hypotheses".[12] This position is, according to Delacre, based on a down-to-earth realism: "In this small book, we have insisted on confronting the bombastic theoretical pretensions of the champions of conceptions and formulae with a truly down- to-earth method. First come the solid and the undeniable, and only afterwards the imagination, the fantastic, conceptions and conventions."[13] As Delacre's credo succinctly summed it up: "Only one thing is true, the simple and brute fact".[14] It was books like these — often written by French chemists — that helped to disseminate the myth of chemistry as a science composed only of facts, completely shriven of all theory.

The Instrumentalist Response

We turn now to another French philosopher of science, Pierre Duhem, who, along with Auguste Comte, is much better known in the US than the others in the tradition of French positivism whom we have been discussing. Duhem, like other positivists, regarded scientific theories as conventional. For him, they were simply instruments for the organization and classification of data, which, while they could be better or worse, did not denote any underlying reality. Duhem emphasized the importance of experimental data, but was not naïve enough to believe that it was possible to speak of empirical data completely independent of any theory, as we saw in Chapter 9.[15] Furthermore, due to a longstanding personal animosity between the two men, Duhem was in no hurry to join the ranks of Berthelot's supporters. As might be expected of someone adopting an "instrumentalist" position, that is someone who regarded theories as instruments rather than descriptions of reality, Duhem remained sceptical about atomism. This resistance to the atom went further than simple scepticism, however, as he actively sought a way of interpreting atomicity that would dispense with the need not only for atomic formulae, but also for a plurality of elements. He believed that there was another approach that could allow chemists to describe and predict chemical reactions while dispensing with molecular formulae, which, despite all the precautions and sceptical convictions of the chemists, perpetually threatened to turn them into naïve realists. Due to his antipathy towards Berthelot, Duhem instead cited Henri Sainte-Claire Deville in support of his anti-realist position:

> Every time someone has tried to imagine or to draw atoms or the groups of atoms in molecules, I believe they have done nothing other than offer an inept reproduction of a preconceived idea, of a gratuitous hypothesis, of a sterile conjecture. These representations have never inspired a serious experiment, and never serve to prove, but only to seduce.[16]

Despite his references to this anti-atomist movement, Duhem did not himself go as far as rejecting the distinction between atoms and molecules. As far as structural formulae were concerned, while he acknowledged that they could be useful, he nevertheless regarded them with deep suspicion because they could lead chemists into the error of thinking that chemical substances remained essentially unchanged when they entered into chemical combination.[17] "The chemical formula in no way expresses what really persists in the compound but rather that which is potentially there, that which can be extracted by the appropriate reactions." Thus, the organic chemists' attachment to their representations of molecular architecture prevents them from responding to Aristotle's fundamental question: what is the mode of existence of the elements that enter into a mixt?

Duhem chose to respond to this question by way of physical chemistry. His project certainly did not lack ambition, as he aimed to dispense with any kind of molecular imagery and base his analysis solely on the observation and measurement of chemical phenomena. This meant considering such empirical events as the disappearance of one or more bodies, the appearance of others bearing different properties, volume changes respecting the conservation of mass and the emission or absorption of heat. Duhem intended to consider only measurable properties, and took the notion of state (and change of state) as the key element in his new theoretical reconstruction of chemistry. He translated the Aristotelian notion of power or potential into thermodynamic potential, an abstract notion that would allow him to define a state in mathematical terms without describing it, and above all, without having to pronounce on the nature of the "matter" in the system undergoing the transformation. Duhem intended to use mathematical expressions of the potentials of the reactants as the means to determine the proportions of substances entering into chemical combination without the "fiction" of atoms. He believed that this approach would avoid having chemists imagine that the reactants were actually present in the compound formed by their reaction.

Once we have understood this approach, we can see why Duhem considered chemistry the heir to Aristotle's philosophy. By means of a qualitative physics, he aimed to describe the transformation inherent in the formation of a mixt. He could thereby save the chemical phenomena while avoiding what was for him a naïve realist position associated with thinking in terms of atoms and molecules. The paradoxical result was that Duhem had succeeded in depriving qualities of their substance by reducing each one to a numerical value that served to represent it in his theoretical construct. By removing the ontological basis of the Aristotelian notion of potential or power, he had succeeded in constructing a theory capable of representing chemical reactions in purely mathematical terms, an approach he named "chemical mechanics". The theory was also capable, in principle at least, of predicting the circumstances in which reactions would take place, and all of this without making any assumptions concerning the mode of existence of the participating elements during the chemical transformation. Using quantitative measures of observable effects combined with abstract formulae, it is possible to establish equalities and differences without identifying what is being measured, thereby creating a mode of theorizing about chemical reactions that is completely abstracted from the nature of the entities entering into them.

This approach leads to another paradox, this time a disciplinary one. For Duhem, the best way to treat the central issue of the mixt in chemistry was to deploy the theories of physics. While Duhem was inspired by thermodynamics and not by the more classical Cartesian mechanical physics or Newtonian dynamics, it was nevertheless a form of mathematical physics that allowed him to avoid any question concerning the "reality" that might lie behind the observable phenomena that supply the data. For Duhem, therefore, chemistry's philosophically purified future lay in thermodynamics, which explains why his book on the history of chemistry closes with a chapter on the history of thermodynamics.

The Energetist Response

Wilhelm Ostwald also tried to resist the temptation of realism applied to atoms. To this end, he deployed a version of physical chemistry characterized by an extreme mathematical abstraction. Here again, measurements were not conceived in terms of comparing objects, but simply by the comparison of numerical values. In contrast to Duhem, however, Ostwald was interested in the kinetics of chemical reactions, with his work on catalysis earning him the 1909 Nobel Prize for chemistry. Ostwald decided not to work on completed reactions, and turned instead to an area formerly studied by Duhem himself, slow equilibrium reactions between acids, bases and salts in solution, where the outcome is an equilibrium between the reactants and the products. In fact, various antagonistic reactions are going on simultaneously, arriving at a state where they exactly compensate for one another while continuing unabated. In order to interpret this type of reaction, which, at the beginning of the nineteenth century, Claude-Louis Berthollet had argued was the model for all chemical reactions, chemists transposed Clausius's kinetic hypothesis, originally proposed in 1857 to explain vaporisation. This theory was worked out on the basis of the frequency of collisions between particles, using calculations of mean velocities based on their energies. The equilibrium state was reached when the collisions between molecules (or ions) responsible for the reaction proceeding in one direction exactly balanced those responsible for the reaction going in the opposite direction. Nevertheless, Ostwald aimed to use this theory to describe equilibrium reactions, but without its accompanying atoms and their atomicity. Indeed, he wanted to dispense with any conception of solid material bodies and their interactions that lay at the base of classical mechanics.

In 1895, Ostwald shook the scientific world with his announcement of the "rout of atomism", triggering a violent debate in both Germany and

France. Ostwald's goal was to radically reform both chemistry and physics, dispensing with any concept of matter and admitting only the concept of energy combined with the two fundamental laws of thermodynamics. He dismissed the principle of the conservation of matter as a useless metaphysical residue of an outmoded form of mechanics. He considered it metaphysical because it implied certain hypotheses; chemists suppose that an entity (a substance) is conserved even when all the phenomenological properties disappear, and yet, a reality is only defined by its physical and chemical properties.

Ostwald eventually abandoned his struggle against atomism at the turn of the century when Jean Perrin succeeded in harnessing a powerful set of arguments in favour of the existence of atoms, including a number of considerations based on kinetics.[18] In his book, *Les atomes*, Perrin sought to transform the status of atoms from fiction to fact by determining Avogadro's number in thirteen different ways. Avogadro's number — N — is the number of molecules purported to be in the mass of a compound that is equal to its molecular weight in grams. Thus, to take an example, two grams of hydrogen gas, composed only of the isotope hydrogen (atomic weight of one), should contain N molecules of H_2. Perrin noted that a wide variety of phenomena as diverse as Brownian motion; osmotic pressure; the electrical conductivity of ions in solution; the colour of the sky; the diffusion of light by a gas; the spectra of radiation studied by Max Planck; cathode rays (electrons); and radioactivity can be interpreted in terms of the frequency of events and so translated into approximations of N. While it was clear that no one could see any individual atoms or molecules, there were many ways to count them. Using thirteen different methods to arrive at Avogadro's number, Perrin converged on a single value of approximately 6×10^{23}. This "miracle of concordance" was, Perrin argued, possible only because atoms and molecules really did exist and were responsible for the observable phenomena at the basis of his calculations. Thus, scientists could safely venture beyond the phenomena and try "to explain complex visible things by simple invisible things". Nevertheless, Perrin did not believe that this approach would ultimately simplify the explanation of observable phenomena, as the invisible world was populated by a host of different interacting entities that the chemist had to elucidate at the level of molecular reality.

What conclusions can we draw from our exploration of these various chemical philosophies? The main conclusion is that chemistry has always challenged the longstanding philosophical divide between positivism and realism. While this science has served for some as a model for the elaboration of a philosophical doctrine refusing to address the ontological question of the reality of atoms and molecules, it is rare for any chemist to deny this reality. Meyerson was no doubt right when he supposed that chemists believe wholeheartedly in the atoms and molecules that form the substances they manipulate in the laboratory. Nevertheless, chemical atomism, as we have explained, is not the same as this kind of realism, and corresponds to a need to represent the capacity of atoms and molecules to enter into relationships with others. Thus, while chemists can be labelled as both positivists and realists, it is perhaps better to pose the question of their philosophical allegiances in different terms.

References

1. For a more detailed discussion of chemists' atomism, see A. Rocke (1984), (1993) and (2001).
2. J.-B. Dumas (1837), pp. 196–218. The specificity of chemical atomism is a central concern of Alan Rocke's works cited in the previous footnote.
3. A. Laurent (1837), quoted in J. Jacques (1954).
4. M. Blondel-Mégrelis (1996), pp. 81–82.
5. A. Kekulé (1867). See also B. Görs (1999).
6. G. Urbain (1921), p. 9.
7. E. Meyerson, 'L'évolution de la pensée allemande dans le domaine de la philosophie des sciences' unpublished conference delivered on 23 April 1911 (Meyerson archives A 408/11), p. 22. Meyerson was shocked by this 'reluctant' atomism (*atomisme du bout des lèvres*), which he considered a form of duplicity if not hypocrisy. Elsewhere, he confessed that it was this stance that prompted him to take up the philosophy of science.
8. E. Cassirer (1953).
9. C. A. Wurtz (1868–1878) *Discours préliminaire*, vol. I, pp. LXXV–LXIX.
10. M. Berthelot (1877), p. 1194.
11. C. A. Wurtz (1877), on p. 1268.
12. M. Delacre (1923), p. 157.

13. *Ibid*, p. 158.

14. *Ibid*, p. 13.

15. See footnote 2, Chapter 9.

16. H. Sainte-Claire Deville (1886–1887), p. 52 cited in P. Duhem (1902), p. 151.

17. P. Duhem (1892).

18. J. Perrin (1913).

CHAPTER 12

AGENCY AND RELATIONS

In this chapter, as we head towards the conclusion of the book, we will leave the domain of epistemology in order to consider the chemist's ontology. Having offered a critical evaluation of the power to know things, now we want to try and understand what type of ontology can be adopted by chemists, or, more precisely, what ontology is appropriate to their scientific practice.

Rather than framing the preliminary epistemological debate in terms of the question "what can one know?" it might be better to pose the question "what can one do?" and then examine the ontological consequences. This approach has the advantage of taking the dual nature of chemistry into account, recognizing chemistry as both science and industry, as being essentially productive. Indeed, this was the very question that Aristotle posed in connection with the mixt in his *On Generation and Corruption* (*De Generatione et corruptione*), a fact that Duhem seems to have overlooked in his recuperation of Aristotle's approach. Thus, Aristotle was not so much concerned with the nature of the mixt as with the possibility of mixing materials to produce new ones. Aristotle's interrogation concerned the conditions necessary for the process of a chemical reaction, and so he offered very precise descriptions of what was required to make such mixts.[1] We do not want to suggest that these conditions proposed by Aristotle constitute some sort of proto-chemistry, however, we just want to underline the point that it is impossible to define the status of the object of chemistry if one does not take into account the duality of chemistry as both natural science and productive technology. The chemist's ontology should therefore be considered in relation to a project of intervention or activity.

The trick to understanding the object of chemistry is, therefore, to stop asking to what extent our theories can transcend the level of phenomena to reveal the reality that lies behind, and instead to ask what we can do with these phenomena. What's more, chemistry as heir to ancient practical arts like metallurgy, dyeing, pharmacy, and glassmaking, has never truly abandoned this technological dimension. Even after being promoted to the academic heights of a "pure science", chemistry continued to produce a multitude of applications, although for reasons of status, the distinction between pure and applied had to be constantly reinforced even as it was perpetually being blurred. Thus, chemistry has a characteristic relationship with its objects quite different from the other natural sciences precisely because it has always inhabited the frontier zone between science and technology.

Elements as Actors

The chemist's art consists of managing populations of molecules in order to bring about the desired reactions. In this context, we can reason in terms of elements in so far as these abstract entities can serve as useful working tools.

Following this approach of treating elements as tools, we can consider them to be operational in two different senses. First, they are operational in the classic sense illustrated by Lavoisier's famous definition of the element, which we cited at length in Chapter 9. Here, Lavoisier defined his elements as "all the substances into which we are capable, by any means, to reduce bodies by decomposition. Not that we are entitled to affirm, that these substances we consider as simple may not be compounded of two, or even of a greater number of principles; but, since these principles cannot be separated, or rather since we have not hitherto discovered the means of separating them, they act with regard to us as simple substances…".[2] Lavoisier's use of "with regard to us" and "we" conveys the relativity of this conception of element that depends on the analytical techniques available to the chemical community at any given time. This means that the list of elements is not simply provisional, but depends on these techniques that provide the limits to decomposition. Allowing the possibility that elements can evolve along with the relevant techniques, changes their nature. They are no longer conceived as the ultimate building blocks of nature but are instead bound to laboratory operations. This is the first sense of an operational concept; one that does not seek to pick out the essence of the object under consideration, but instead identifies the operations needed to obtain it.

A concept can also be operational in a different sense. Thus, it can be defined in terms of the operations the designated entity can perform. Lavoisier points to this other definition when he says of his elements that "they act with regard to us as simple substances". What is important for him as a chemist is that the elements are actors in chemical operations, and so are defined by how they act and react in a network of relations with other chemical actors. This second sense of operational definition is more fundamental than the first, and was characteristic of chemistry even before Lavoisier introduced this explicitly pragmatic definition of the element. As early as the eighteenth century, if not before, chemists had adopted the habit of identifying substances by what they were capable of doing. Thus, in his "Essay on the Analysis of Common Sulphur" (*Essai sur l'analyse du souffre commun*) published as an article in 1703, William Homberg showed that the acid derived from the decomposition of sulphur was "perfectly the same thing" as oil of vitriol, precisely because this "spirit of sulphur" could do the same things as oil of vitriol and vice versa.[3] Following this logic, the traditional four elements can be redefined in terms of their action on other bodies, interpreted as various means of action that can be deployed by the chemist just as they can be used by nature. Hermann Boerhaave explicitly presents the four elements as chemistry's "instruments" in the following sense: "In whatever art where one proposes to change a body, one calls instruments those things upon which one can imprint — if one has not already done so — a movement capable of producing a desired change."[4] Guillaume-François Rouelle, as well as some of his students, treated elements as "natural instruments" and presented them in the same chapter as the one that treated the artificial instruments such as the glass and ceramic vessels used for performing reactions.[5] Thus, all bodies become "instrumentalized" by the chemist who regards not only salts, but even affinities themselves as instruments that can be put to use. From this perspective, nature itself is conceived of as a vast laboratory, a place, like the chemist's own laboratory, where bodies are put to work. Indeed, as Larry Holmes so astutely remarked, eighteenth-century chemists always used the term "operation" to refer to what we would today call a "reaction".[6] The elements were actors in the *opera* — a term that simultaneously referred to the chemical work operated by nature and the work of the chemist (see Figure 3 on page 34 for an allegorical representation of "the" *opus*).

By turning nature into a theatre of operations where they could work in partnership with natural substances, eighteenth-century chemists were once again blurring the boundary between nature and art. Thus, when chemists used the metaphor of nature as a theatre, it was not in the same sense as Fontenelle used it. For Fontenelle, the erudite natural philosopher could expose the mechanism behind the on-stage effects (there to impress the ignorant) by illuminating the functioning of the cogs and wheels in the machinery hidden in the wings or behind the stage.[7] For the chemist, on the other hand, the theatre was meant to evoke not only the variety of actors who performed on stage, but also their constant circulation. This disparity reflects a difference in the approach to matter, as chemists do not look at the material world as an illusory, phenomenal spectacle, nor do they see themselves as passive spectators. Finally, and most controversially, the chemists' principal goal is not to offer a veridical representation of the world.

While this way of defining bodies by their actions was particularly actively pursued by eighteenth-century chemists, it was by no means confined to this period. We have already seen how, following Laurent's example, nineteenth-century atomists conceived of molecular structures not as concatenations of so many elementary building blocks of nature but as chemical agents that could participate in addition or substitution reactions. This thinking explains Laurent's choice to classify the elements according to their function. Thus, despite the obvious differences in their properties, he placed chlorine and hydrogen in the same column because they could be exchanged for one another in substitution reactions. They perform the same function and can be considered equivalent actors performing on the same stage of organic chemistry. Indeed, it was in organic chemistry that the notion of function displayed its full heuristic power, leading to an innovative productive way of grouping and thinking about compounds. How else could one even contemplate organizing around six million organic compounds if not under the twenty or so classes defined by functional groups like the acids, alcohols, aldehydes and ketones.

The evolution of the rules governing chemical nomenclature is quite instructive in this context of the functional definition of chemical individuals and species. Indeed, from this perspective, the reforms in chemical nomenclature championed by Lavoisier and his collaborators appear as a sort of interlude rather than a foundational episode in modern chemistry. The new chemical nomenclature introduced by Guyton de Morveau, Lavoisier, Berthollet and Fourcroy in their *Method of Chemical Nomenclature* of 1787, insisted that the names of compounds should represent the constitutive elements and their relative proportions. Because this principle continues to form the basis of our current system for naming compounds, it seems quite obvious to us, and it is easy to dismiss or at least forget the system that preceded it. Before the reforms of the *Method of Chemical Nomenclature*, many compounds and even elements possessed common or vernacular names that evoked their use in pharmacy or other chemical arts. Thus, for example, "English Laxative Salt" became magnesium sulphate following the reform, and "Soda Hispanica", which evoked its origins in its name, was replaced by the generic term sodium carbonate.

If apothecaries or other chemical artisans wanted to use the new language, it meant giving up these kinds of names that referred to some function performed by the substance, and instead adopting names that indicated only its elementary composition. The new language aimed at offering a faithful representation of the composition of the substances to be named. As Lavoisier put it in his introduction to the *Method*; "The perfection of the nomenclature for chemistry [...] consists in rendering the ideas and the facts in their exactitude, without suppressing anything that they present and — above all — without adding anything. It should be nothing other than a faithful mirror, because, as we cannot repeat too often, it is never Nature nor the facts she presents that betray us, but our own reasoning that does so."[8] While the principle that the name should represent the elementary contents of a molecule underwrites official nomenclature, the rise of the structural formula represents an acknowledgement that a simple concatenation of constitutive elements alone is insufficient to account for the nature of chemical compounds. Furthermore, when the first graphic representations of molecular structure were introduced in the nineteenth century, the goal was not so much to represent the real architecture of molecules as to classify the existing ones and predict the existence of new ones. Indeed, to use a term developed by Ursula Klein to talk about the link between representation and practice in chemistry, these formulae functioned as "paper tools".[9] Just like the molecular models constructed out of balls and rods, which were introduced by Wilhelm Hofmann in the nineteenth century, these representations could be manipulated and displaced. This representation allowed the chemist to imagine, or more accurately, to visualize the different isomers of a compound that can be obtained by the displacement of a double bond or some other transformation. In this context, we can recall that van't Hoff introduced the tetrahedron form of the carbon atom, which inaugurated the era of three-dimensional formulae, as a way to account for the experimental result that substituted alkanes do not exhibit as many isomers as a planar structure would lead one to expect.[10] Whether planar or three dimensional, the primary purpose of such chemical formulae is not to represent the structure of molecules, but to predict behaviour and to help in the construction of new molecules.[11] Thus, in organic chemistry at least, the name says less about the intrinsic nature of the molecule in question than how chemists perceive it, in particular the use they can put it to, and what useful performances they can get out of it.

Operational Realism

What we are trying to suggest here is that chemists' practical philosophy is not limited to their notorious scepticism. The fact that they can marshal a population of material entities to perform useful work is itself a distinctive mark of chemists' conceptual practices. For chemists, invisible entities are not primarily the keys for understanding the material world, the reality behind the phenomena, but rather a set of tools or instruments with which one can bring things about by acting in the world.

In this interpretation, therefore, action comes first, before conceptualization, nomenclature, or theory. It is tempting to qualify this philosophical position as "instrumentalism" but this term is already applied to the philosophical position that interprets theories as conventional tools for calculation or classification without making any claims concerning the reality of the theoretical entities they deploy. This classic instrumentalism is an anti-realist (or at least anti-scientific-realist) stance that treats its concepts as constructs of the human mind without any referent in the real world.

Chemists, on the other hand, rarely question the reality of the tools with which they do their chemical work, be they natural or artificial. In this sense, Meyerson was completely justified in condemning the duplicity of chemists like Kekulé. These nineteenth-century chemists avoided the metaphysical question of the existence of atoms while they happily continued to use them, like a plumber would use a spanner or a screwdriver. In his 1983 introduction to the philosophy of science, *Representing and Intervening*, Ian Hacking makes the crucial distinction between "realism with respect to theories" and "realism with respect to entities". Chemists' realism is an entity realism, although one that is best qualified as "operational realism" because of its perpetual link to productive activity.

In order to understand the type of objects that chemists deal with, we need to put aside the debates that dominate contemporary philosophy of science, in particular, the question of whether or not science represents reality. The chemists' realism is more like the "entity realism" that Hacking constructs around the example of the electron. It is, he explains, the fact that one can manipulate these electrons in a reliable predictable manner to generate effects that convinces the scientist (and the philosopher) of their reality.

> Experimental work provides the strongest evidence for scientific realism. This is not because we test hypotheses about entities. It is because entities that in principle cannot be "observed" are regularly manipulated to produce new phenomena and to investigate other aspects of nature. They are tools, instruments not for thinking but for doing.[12]

According to Hacking, what is important is not the use of electrons as theoretical hypotheses in physical theory to explain or save the phenomena, but rather their practical deployment in experimental settings to create phenomena and, more widely, their ability to influence the outcome of observable events.[13] Experimental scientists, the argument goes, are spontaneously realists not because they believe in any transcendent theory of matter, but because they can do things with the putative invisible entities in question. What convinces experimenters that such objects exist is that they can use their experimental apparatus to "spray" or "burn off" electrons, thereby generating predictable observable results. The argument that Hacking makes concerning electrons applies equally well to the entities put forward by chemists over the course of time, such as the element/principle of the eighteenth century, Mendeleev's concept of the element from the nineteenth, or atoms and electrons in twentieth-century chemistry.

Adopting this philosophical stance, we can now understand why none of these concepts that have appeared in the history of chemistry can ever become truly obsolete. The concept of the element/principle, popular before the rise of Lavoisier's chemistry is still valid in so far as the idea of a material entity that bears certain properties or is capable of triggering a specific behaviour is pertinent. Likewise, as we saw when we considered the origins of the periodic table, the concept of the element developed by Mendeleev still makes sense if one conceives of this table as a useful organizational scheme for the elementary tools available to the chemist. Nevertheless, this reference to Mendeleev's concept of the element complicates the notion of operational realism that we have been presenting, because, unlike the electron around which Hacking constructed his entity realism, these elements are abstractions and not straightforward entities. One cannot manipulate Mendeleev's elements in the same way that one can spray electrons, even with such advanced technology as the scanning tunnelling microscope at the base of so much nanotechnology. It is, however, important to consider these different kinds of "objects", as they constitute an important part of the practical resources of modern chemistry.

Limiting their ontology to agents that can act would be insufficient for chemists; they also need to postulate capacities for action. Thus, the Aristotelian notion of *dynamis* — meaning power or potential — is no more redundant than the ancient concept of the element/principle. Although this idea of *dynamis* does not correspond to the Cartesian criteria for true ideas — that they should be clear and distinct — the principle of potential or a capacity to act is something that it would be very hard for chemists to do without. It is important to note that our insistence on the importance of this concept takes us out of the paradigm of efficient causes (mechanical interaction or the action of forces on objects) that has shaped modern physics. This expansion of the chemist's realism to include capacities or potentials makes it more complex than the version of entity realism presented by Hacking. This position is closer to the one defended by Nancy Cartwright, who has consistently argued for the reality of nature's potential, and has tried to show how the recognition of such capacities is implicit in any scientist's understanding of a general law of nature.[14]

It should, by now, be clear to the reader that the popular image of chemistry as a superficial empirical science obliged to seek its philosophical foundations in other more fundamental sciences is quite inaccurate, if not philosophically defamatory. Whether this vision of chemistry is the deliberate construction of philosophers of science with a predilection for physics, or just results from the lack of attention paid to chemists' concepts and methods, it does a great disservice to philosophy, depriving it of an interesting practice-based approach. Chemists have adopted an anti-essentialist perspective on the material world, regarding it as populated by individuals with a range of capacities to put themselves in relation with one another, thereby producing the phenomena observed in the laboratory and the chemical world. The chemist does not adopt the ancient philosophical ideal of going beyond the appearances to attain a hidden reality, but instead remains at the level of chemical phenomena which are too fascinating to be put aside. Indeed, the corresponding philosophical position is that there is no ultimate hidden reality behind the phenomena. There is no mechanical puppet-master behind the scenes that explains the phenomenal world; all there is to the world are material agents that allow new properties to emerge through their different relationships with one another.

Some might surmise from the argument we have been presenting that the chemist's ontology is less rich or less serious than that of the other sciences, in particular physics, but this would be a wrongheaded conclusion. Chemists are realists in the strongest sense of the term, with a solid belief in the reality of the actors that they put to work, be they electrons, elements, alcohols, or ions. This should not, however, be confused with the substantialist realism that Bachelard was so quick to condemn. Instead, the chemist's realism is the robust faith that Meyerson attributed to practicing scientists. This realism, as we have been arguing, does not derive from some innate tendency to objectify the concepts with which one works, but instead reflects the intimate relationship between manipulation and realism. This intimate conviction concerning the reality of the entities manipulated in laboratory experiments is not, however, unique to chemists but is shared by all experimental scientists and technicians. What distinguishes chemistry from the other sciences is that the entities put to work in this field exist not only in the mode of actuality, but also in the complementary mode of potentiality. The causal power the chemist attributes to such material abstractions both renders the science predictive and allows the formulation of general laws.

Alternative Metaphors

Philosophers might be tempted to criticize chemists' operational realism as a form of naïve credulity that fails to measure up to a purer form of scientific realism concerned with the reality behind the phenomena. This criticism, however, assumes that the aim of science is or at least should be to represent an external, independent reality. To better understand the difference between these positions, it is helpful to consider the metaphor of the three blind men investigating an elephant using their hands. What object do they have before them? The first, feeling one of the elephant's legs declares that it is a tree trunk, and the object is a tree. The second, after running his hands over the trunk announces that the object is a snake. The third, feeling the elephant's tail suggests that the object is some sort of whisk to keep the flies away.[15] The way this metaphor is presented serves well to illustrate how the problem of realism is posed within standard epistemology. First of all, in setting the scene, one assumes that there is some reality — some sort of well-defined object (in this case the elephant) external to the human knower, and that the objective of the investigator's method is to arrive at the most accurate possible representation of this reality. Within this context, the metaphor suggests that each blind man (usually taken to represent the different scientific domains or disciplines) has only a partial grasp of the whole that he projects as the reality. Nevertheless, the implicit message of this metaphor is that combining these partial visions will lead to the truth, where the truth is conceived of as an accurate description of the elephant that is *really* there and is responsible for each one of the partial perceptions experienced by the individual investigators.

Nevertheless, it does not take a great deal of philosophical sophistication to see that this metaphor of the elephant is constructed around a highly contentious vision of reality. Just as the blind man needs to step out of his restricted perceptual condition to confront the reality behind his experience (to be able to see the whole elephant), so scientists, in order to be sure that there is a coherent reality behind the results of their experiments or observations, need to step out of their condition as situated knowers — the human condition — a move impossible for any human being to make.

The chemists' operational realism that we have been arguing for does not postulate any underlying coherent reality, and as a corollary their goal is not the faithful representation of such a reality. Fortunately, there is another metaphor involving elephants that offers a much more adequate description of the chemist's engagement with the material world. Primo Levi, the Italian chemist and concentration-camp survivor, uses this metaphor to describe the chemist's art in his autobiographical work, *The Monkey's Wrench*. Here, Levi compares chemists to elephants struggling to create jewellery in a goldsmith's workshop. Evidently, they have to put in a great deal of effort for a small return, and there are many creations that they imagine to be possible but are unable to realize. This situation differs dramatically from Plato's myth of the cave in which humans are limited to a blurred, degraded vision of a transcendent world of ideas. For the elephants in the goldsmith's, the limits are imposed by an inherent clumsiness in the manner of coming to grips with the world. There is an approximation introduced by a play of scale, which gives rise to a mismatch. The mismatch is not, however, between the tools of the goldsmith and the elephants' feet (as this would be to assume that the tools were designed for a certain size of hand, which while true in this metaphor is not the case for the chemist), but between the conception of what might be possible and what the elephants are able to achieve in the circumstances. Chemists practice their art in a space of possibilities constrained by the innumerable relationships that exist between all the various actors implicated in the reactions, homogeneous and heterogeneous alike. The chemists' reality is not, therefore, a posited ultimate explanatory mechanism or ontology underwriting the observed phenomena. A better understanding of the chemists' reality would come from comparing it to the shared basis of a discussion among a group of elephants who are attempting to coordinate their practical efforts in pursuit of some technical or cognitive project in a goldsmith's workshop.

Another thing to note about chemists' realism is its irreducible multiplicity, which reflects the diversity of their tools. Rather than pursuing the holy grail of physics and its philosophy (*the* ultimate reality captured by *the* unified theory of nature), the chemist needs to postulate and consider a huge diversity of material entities each with its own particular constraints. This characteristic diversity means that the debates between chemists over the defining object of chemistry — the molecule, the element, the atom, etc. — are bound to be inconclusive and ultimately fruitless. Thus, chemistry lies between the two venerable scientific traditions of physics and natural history. While the former seeks to establish the properties of matter in general and the ultimate mechanisms that underwrite them, the latter adopts a descriptive approach. Indeed, there is a clear natural historical streak in chemistry, with endless monographs describing the particular properties of compounds or reactions, a literature that chemists themselves refer to as "zoology". But all is not "stamp collecting," chemistry also boasts a considerable body of theoretical knowledge, including its accounts of chemical bonding and reaction mechanisms, which equally has no ambition to represent the ultimate structure of the material world. This double nature of the science, with its descriptive and theoretical aspects constitutes a major feature of the epistemology of chemistry.[16]

One reason to argue for the existence and pertinence of a characteristic philosophical perspective associated with chemistry is that it can put an end to the enterprise of reducing all the sciences to a single, fundamental one from which all the others can be derived. Accepting a legitimate philosophy for chemists has the added advantage of acknowledging the multiplicity of possible modes of existence, or, more concretely, of leaving the way open for different modes of engaging with reality. Scientists will not behave in exactly the same way if they are confronted by what they perceive to be uniform, passive, brute matter or by bodies made up of "intelligent" matter, be it active or endowed with the capacity to act. Many contemporary critics of science complain that science is soulless, interested only in the interaction of passive machines, but if they knew chemistry better, they might think differently.

References

1. First of all, one needs both active and passive elements that can alter one another in a reciprocal manner to produce a new homogeneous body. Thus, totally passive or inert bodies cannot be mixed together. Second, the bodies need to be divisible into fine particles, which is why liquids mix together so easily. Third, in order to have a truly novel body emerge, one needs to ensure that the proportions of the mixt's constituents are approximately equal. Thus, anyone can see that pouring a drop of wine into a large body of water fails to produce anything different from water.

2. A.-L. Lavoisier (1789), Transl. R. Kerr, p. xxiv.

3. W. Homberg (1703).

4. H. Boerhaave (1745) French transl. vol. 1, p. 267.

5. In his *Institutions chymiques* Rousseau presents the four elements in the following terms: 'In order to establish an artificial laboratory on the model of Nature's laboratory, one cannot simply consider the general way in which she works, one needs above all to know perfectly the instruments she uses. There are a great number of these instruments: the sun, water, salts, earths, even the parts of bodies variously put in motion and figured. But one can reduce all these to four general classes — water, fire, earth, air — by means of which all natural bodies exist, are produced, and are conserved or change in accordance with the laws established at the very beginning.' J.-J. Rousseau (n.d.), p. 63.

6. F. L. Holmes (1995) and (1996).

7. B. Fontenelle (1686).

8. Antoine-Laurent Lavoisier, 'Mémoire sur la nécessité de réformer et de perfectionner la nomenclature de la chimie, lu à l'Assemblée publique de l'Académie royale des sciences du 18 avril 1787', in L. B. Guyton de Morveau et al. (1787), p. 69.

9. U. Klein ed. (2001), pp. 13–34.

10. P. Ramberg (2001).

11. International efforts to standardize the nomenclature of organic compounds have tended to go in the same direction. The rules fixed at the Geneva Congress in 1892 demanded that the official name express the structure, giving priority to the longest carbon chain and using suffixes to indicate the functional

groups, and prefixes for the substitution groups. The Liège Congress in 1930 issued rules privileging the functional groups. The name also has to indicate the position of any double or triple bonds, as these are the molecules' principal sites of reactivity. See F. Dagognet (1969), pp. 176–177 and B. Bensaude-Vincent (2003).

12. I. Hacking (1983), p. 262.
13. *Ibid*, in particular chapter 16, 'Experimentation and scientific realism', pp. 262–275.
14. N. Cartwright (1989).
15. See the entry 'réalité' in I. Stengers and B. Bensaude-Vincent (2003).
16. This dual nature of chemistry is the main feature that Linus Pauling sought to convey to college students with his successful textbook; L. Pauling (1950).

CHAPTER 13

TAMING THE NANOWORLD

To Eric K. Drexler, the champion of molecular manufacturing, the traditional art of chemical synthesis looks like a dirty and very primitive way of making things. The chemist chooses certain reactants that are then mixed up together in a vessel in the hope that a sufficient number of molecules will eventually fall into the right place to make the desired product. Drexler advocates a radically new technology that will manipulate individual atoms and molecules, piecing them together like Lego® bricks, and thus providing complex molecular products cleanly and efficiently. In contrast to this dream of synthesis without waste, Drexler characterizes current organic synthesis as a haphazard and somewhat unreliable process for making complex molecular chains.

> Chemists have no direct control over the tumbling motions of molecules in a liquid, and so the molecules are free to react in any way they can, depending on how they bump together. Yet chemists nonetheless coax reacting molecules to form regular structures such as cubic and dodecahedral molecules, and to form unlikely-seeming structures such as molecular rings with highly strained bonds. Molecular machines will have greater versatility in bondmaking, because they can use molecular motions to make bonds, but can guide these motions in ways that chemists cannot.[1]

In *Engines of Creation*, published in 1986, Drexler contrasts two styles of technology. On one hand, current technology handles the building blocks of matter in bulk, while on the other, the coming era of nanotechnology "will handle individual atoms and molecules with control and precision".[2] Chemical synthesis belongs to the ancient tradition of bulk technology dealing with billions of atoms at a time that was initiated by chipping flint and is still used for making microcircuits. Drexler offers a metaphor intended to illustrate the contrast between this kind of top-down method of chemical design and the new bottom-up approach. While the new process is similar to the construction of an automobile piece by piece, the traditional art of synthesis is like trying to make one by taking all the parts at once and shaking them up in a box in the hope of getting a working machine at the end of the process. Drexler concludes that: "It is amazing that chemists are able to do anything at all, and in fact, they have impressive and growing accomplishments".[3]

Although this remark could be interpreted as praise for the synthetic chemists' skills and ingenuity, Drexler instead uses it to criticize their lack of control and the messiness of their approach. Thus, while chemists rely on the haphazard motions of large numbers of molecules in a liquid, in Drexler's ideal molecular production plant, nanorobots will select and place individual atoms to assemble larger molecules. Nanotechnology is seen to be similar to genetic engineering in its projected use of molecular machines to perform specific tasks, while synthetic chemists are unable to control their chemical reactions with any great precision. As a result, molecular manufacture will provide a clean and environmentally-friendly chemical industry that will replace the dirty, polluting chemical plant. Furthermore, while these traditional chemical factories expose the population to a range of hazards associated with unstable or otherwise dangerous reagents and intermediate products, molecular manufacturing will be entirely safe, and with no undesirable by-products.

Indeed, Drexler takes genetic engineering as his model for nanotechnology, inspired by the fact that molecular biologists describe ribosomes and proteins as molecular machines. The hope is that nanotechnology will be able to construct similar nanomachines programmed to perform specific industrially useful molecular tasks. The major difference between these projected nano-machines and the natural "machines" operating in cells is that the second are self-assembling. The aim is for tomorrow's nanoengineers to design artificial nanomachines, inspired by ribosomes and proteins that will be able to reliably assemble molecular components. These "universal assemblers" will operate like automated machine tools, bonding individual atoms and molecules with perfect precision. Furthermore, using the resources of the periodic table, these machines will be made of more resilient molecular material than the all too vulnerable molecular machines of the living cell.[4]

Does this logic mean that the coming era of nanotechnology will spell the end of chemistry?

Bottom-Up versus Top-Down

This question has to be taken seriously, as nanotechnology seems to challenge not only traditional synthetic techniques but also the very discipline of chemistry itself. Nanotechnology is characterized as:

> Working at the atomic, molecular, supra-molecular levels, in the length scale of approximately 1–100 nm range, in order to understand, create and use materials, devices and systems with fundamentally new properties and functions because of their small structure.[5]

Three major features distinguish nanotechnology from chemistry. (i) At the scale of the nanometer (10^{-9} meter), it is possible to visualize and address a single molecule rather than operating at the level of N (Avogadro number) molecules; (ii) at this scale, the frontier between inorganic and organic matter no longer makes any sense, and nano- and biotechnology work together; and (iii) molecules, macromolecules as well as genes and proteins, are viewed as machines performing specific tasks rather than as building blocks of matter.

Perhaps more than any other, the field of nanotechnology instantiates what has been described as a new regime of knowledge production oriented towards industry. In this context, it is impossible to distinguish "pure" and "applied" research.[6] Characterized by its high degree of interdisciplinarity, research in nanobiotechnology also blurs the boundaries between traditional academic disciplines such as physics, chemistry and biology. Various interdisciplinary configurations are now being developed, ranging from molecular genetics to synthetic biology, which may deeply affect chemistry's identity and may even bring about the end of chemistry as a separate discipline. So let us now consider what chemists have to say about the future of chemistry.

Drexler's concept of molecular manufacture has been submitted to merciless criticism from a number of chemists. Richard Smalley,

George Whitesides, and others have argued that Drexler's molecular assembler capable of moving parts to the right position for assembly is chemically impossible.[7] Smalley has raised two objections: not only would "molecular fingers" obviously take up too much space and lack the precision needed to conduct reactions at the nanoscale (the "fat fingers" problem), but they would also adhere to the atom being moved, making it impossible to move a building block where you want it to go (the "sticky fingers" problem). In addition, chemists have pointed out that Drexler's nano-manufacture substitutes a kind of Meccano-synthesis for genuine chemical synthesis. Drexler imagines molecules as non-interacting rigid building blocks that can be assembled like the ice-blocks of an igloo. Furthermore, the functions performed by the various parts of the molecular machinery are conceived of as being essentially mechanical. They position, move, transmit forces, carry, hold, store, etc. and the assembly process itself requires positioning the components with the precision associated with mechanical construction. Critics have clearly shown that Drexler's model of a machine is inappropriate for the nanolevel, because he simply transferred a macroscopic machine model without taking into account the radically different environment of the nanoworld. As George Whitesides has pointed out, a nanoscale submarine would be impracticable because of Brownian motion, which would make it impossible to guide.[8] Furthermore, the biomachinery that inspired Drexler does not deal with rigid building blocks either. From the side of biology, Richard Jones has responded that Drexler has misunderstood the nature of the "machines" found inside cells, which are "soft machines". Thus, he indicates three major differences between these bio-machines and conventional human technologies: (a) Instead of channelling the material traffic by means of tubes and pipes, living systems take advantage of Brownian motion to move molecules around. (b) Living systems do not use rigid molecules like the molecules of the synthetic chemist, as proteins can readily change their shape and conformation. (c) The constraints in building machines at the molecular level differ from those of "bulk technology". Inertia is no longer a crucial parameter. Instead, surface forces, viscosity in particular, become major constraints determining whether or not nano-objects will stick together.[9] Philip Ball has suggested that chemistry might provide a better alternative for thinking about nanotechnology than Drexler's mechanical approach.

> I feel that the literal down-sizing of mechanical engineering popularized by nanotechnologists such as Eric Drexler — whereby every nanoscale device is fabricated from hard moving parts, cogs, bearings, pistons and camshafts — fails to acknowledge that there may be better, more inventive ways of engineering at this scale, ways that take advantage of the opportunities that chemistry and intermolecular interactions offer.[10]

If Philip Ball is right, then rather than being left behind, chemists should be at the forefront of this interdisciplinary science. But what possibilities do chemists have for taming the nanoworld? And how exactly are they qualified for this work of molecular engineering?

Rational Design

One response lies in the recent evolution of synthetic chemistry. Like many other domains, the practices of synthesis have been deeply transformed by the use of computers. Twentieth-century chemists, material scientists and pharmaceutical chemists have developed a variety of computer-assisted methods for designing molecules with interesting medical, magnetic, optical, or electronic properties. These approaches are often referred to globally as "rational design" techniques in contrast to the empirical, more haphazard processes of synthesis used in the past.[11] Today's chemist can choose from a wide range of algorithms that use computation, combination, and randomisation to design molecules.

Computational chemistry, which operationalizes quantum theory by means of digital computation, appeared after the Second World War, taking advantage of the machines developed to break codes and support research around the atom bomb. Computational chemistry was initially basic research that remained close to physics. Researchers aimed to build up chemicals *ab initio*, using computers to calculate what was possible, starting from the most fundamental information about atoms and the basic rules of physics. In this respect, the original form of computational chemistry was close to nanotechnology in that it adopted a bottom-up approach. But computers were also being deployed to run models for the molecular mechanics of large systems associated with industrial processes. In the early 1970s, Cyrus Leventhal developed a technique for digitally modelling chemical behaviour based on X-ray crystallographic models in the context of the Multiple Access Computer (MAC) program at MIT.[12] This approach offers a way of avoiding the cost of synthesis by finding out how well a potential compound will work by modelling its chemistry on a computer. Three different perspectives are explored in this approach: thermodynamic features, electronic properties and the spatial, molecular conformation. By visualizing the three-dimensional structure of a compound and rotating it, one can predict how a small molecule might interact with a protein. In this approach, researchers have not limited the use of molecular graphics to visualizing these virtual molecules, but have also developed ways of manipulating them.

Combinatorial chemistry is another computer-based method developed within the pharmaceutical and chemical industries as a cheaper way of creating and identifying potentially useful substances. It consists in reacting together a set of starting materials to generate all the possible combinations and then identifying which of these products might be interesting.[13] In this case, the computer is not used to avoid the "dirty work" of the synthetic reactions and the production of molecules as it is in computational chemistry. On the contrary, the process starts with the synthesis of arrays of related materials with different compositions, the goal at this point being to produce these compounds quickly in small quantities. Once a general, simple synthetic pathway has been selected and optimized, thousands of compounds are synthesized before being screened for specific interesting properties. The idea is to obtain a "library" of substances containing molecules matched with every identified protein target, realizing the maximum possible diversity without any redundancy. Then, with the help of "evolutionary algorithms" run on computers, researchers are able to select the structure that best fits their molecular goal. Thus, this process represents an integration of empirical experimental results — usually from automated processes — with computational techniques.

This method is a form of "rational" design because it is an attempt to eliminate the serendipity of conventional screening methods thanks to the application of the rules of combinatorial mathematics and algorithms of selection. For a number of chemists, combinatorial chemistry is a contemptible method of fabricating substances, and Pierre Laszlo has talked of "the moronic travesty of scientific research known as combinatorial chemistry". For Laszlo, it is a "perversion" of scientific chemistry pitifully limited to one single goal, "the proliferation of chemicals".[14]

However, combinatorial chemistry is more than just a cheap and fast way of designing molecules to be used as drugs or for some other commercial purpose. It is also an exploratory method somewhat similar to the chemical practices of the eighteenth-century chemists who prepared affinity tables. These chemists performed many hundreds of reactions in order to put together their tables, thereby once again illustrating the "natural history" tradition of chemistry. Indeed, while these affinity tables might be considered akin to the modern-day reference literature, libraries of molecules could be said to be the modern equivalent of natural history collections, which often contained the chemicals described in the pharmacopoeia as well as many other more or less exotic species.

Bio-Inspired Chemistry

Another possible response to the challenge of the future of nanotechnology is to turn to the living world for ideas about what lines of research to pursue, which leads us into a domain that has been dubbed "bio-inspiration". Once again, we can ask whether this field, which involves at the very least an intimate alliance between chemistry and biology, threatens the disciplinary identity of chemistry.

The relationship between chemistry and the life sciences, like the relationship it has had with physics, has been an important factor in shaping the public image of the science across its history. In Chapter 3, we noted how the figure of the chemist working to create life in the laboratory survived the collapse of the alchemical tradition and resurfaced once again with the rise of organic synthesis in the nineteenth century. In the early twenty-first century, this image seems to be more present than ever before.

Ironically, it was when the culture of synthetic chemicals was reaching its climax in the "plastics era" of the 1970s and 1980s that chemists started to turn their attention to natural products. Nature, or more precisely living organisms, came back into the world of synthetic chemistry in the context of two different projects. First, living creatures were analysed for their potential use as the source of raw materials to produce environmentally friendly materials. Thus, chemists tried to synthesize biopolymers out of vegetable fibres to make biodegradable rubbish bags and other consumer goods. Second, living organisms served as a source of inspiration in the quest to design high performance, multi-functional composites. Faced with the inadequacy of certain synthetic compounds, materials scientists and chemical engineers realized that better materials already existed in living organisms.[15] Thus, analysing sea shells or the skeletal structure of the most banal insects, researchers started to see how these living creatures naturally produced adaptive structures composed of materials exemplifying optimal combinations of properties given the constraints of the organism's environment. Sea-urchin or abalone shells are amazing bio-mineral structures made out of a common raw material — calcium carbonate. These shells present a complex morphology capable of assuming a variety of functions. Similarly, the silk produced by spiders is an extremely thin and robust fibre with a strength-to-weight ratio that is unrivalled by artificial materials. Wood — the archetypal material — has now been redefined by scientists not only as a composite material made out of long, orientated fibres immersed in a light ligneous matrix but also as a complex structure with different levels of organization that can be observed at different scales. Thus, Nature seems to have already provided elegant solutions to many of the most challenging problems facing modern chemists. As the materials scientist Stephen Mann has optimistically expressed it:

> We can be encouraged by the knowledge that a set of solutions has been worked out in the biological domain. The challenge then is to elucidate these biological strategies, test them in vitro, and to apply them with suitable modification, to relevant fields of academic and technological inquiry.[16]

These biomimetic strategies have prompted new collaborations between biologists and chemists, often under the umbrella of a new interdisciplinary field called Materials Science and Engineering. Biomaterials have taught chemists many lessons: first, most of these materials are multifunctional and represent a good if not maximal compromise between these various functions.

Second, unlike artificial chemical products, biomaterials do not exclude or even avoid the presence of impurities, flaws, mixtures and composites. Third, the examination of their fine structure reveals that biomaterials present a complex hierarchy of structures with different structural features appearing at different levels of magnification.

Bio-inspired chemistry is not, however, confined to attempts to mimic the exquisite hybrid structures of biomaterials. The advent of nanotechnology has also focussed the attention of chemists on the role of biological materials in constructing the substances found in organisms. When it comes to designing substances at the nanoscale, human hands are completely useless, as are all the other tools normally used by the chemist. Drexler conceived of an appropriate tool to overcome this practical difficulty and then coined the term "universal assembler" to describe it. No such artificial "universal assembler" exists, but even the most humble single-cell organism is full of natural, albeit specific assemblers. Furthermore, living cells have found an even more elegant solution for the execution of this nano-synthesis since the assemblers can also assemble and maintain themselves. Self-assembly is a ubiquitous phenomenon in living systems, and it is extremely advantageous from a technological point of view because it is a spontaneous and reversible process with a wide range of applications that generates little or no waste.

Two very different strategies have evolved based on the mechanisms of self assembly found in the living cell. First, scientists can now take advantage of the mechanisms selected by biological evolution to put together the building blocks of living systems, and orientate them to achieve other ends. For instance, it is routine practice today in a number of laboratories to use the complementary structure of pairs of DNA strands in order to produce nanotransistors or other circuits based on a DNA template. In this strategy, chemistry cedes its primacy to genetic engineering, as it is the recombinant DNA that does the synthetic work. Bio-computing is a new field of research that has profited from the potential of DNA to make structures at the nanoscale. By recombining DNA, bio-engineers can use it as a template for making new structures that they can control in detail by using atomic-force microscopy.[17]

The alternative strategy is more strictly chemical, and involves mimicking the biological processes of self-assembly observed in living organisms using the thermodynamic and chemical properties of atoms and molecules. The challenge that chemists face in this context is how to achieve the self-assembly of their components and control the resulting morphogenesis without relying on the system of the genetic code of DNA. To meet this challenge, chemists mobilize all the resources of physics and chemistry during the synthesis. These include chemical transformations in spatially restricted reaction fields, external solicitations (like the use of gravitational, electric or magnetic fields), mechanical stress, and variations in the flux of reagents. They also favour the use of weak bonds — hydrogen bonds, van der Waals forces, etc. — over the making and breaking of covalent bonds between atoms.

Chemists have recently learned other useful lessons from nature, in particular concerning what means to apply to achieve their ends. Over the years, synthetic chemists had grown accustomed to operating with the aid of extreme conditions — high temperatures and low pressures — that are costly in terms of energy, as well as using large quantities of organic solvents that are difficult to dispose of safely once the reaction is completed. Nature teaches us that it is possible to bring about chemical reactions at room temperature and in rather messy, aqueous environments. Imitating these conditions has given rise to a new style of chemistry, for which Jacques Livage invented the name "soft chemistry" (*chimie douce*) in 1977. This approach is used to obtain new materials by performing reactions under quasi-physiological conditions, which generate only renewable, biodegradable by-products. All this at the low cost associated with nature's synthetic processes. The development of soft chemistry has led to the use of increasingly complex raw materials for these reactions, including macromolecules, aggregates and colloids. When combining these large complex molecules, the chemist is no longer limited to considering only the strong covalent bonds between atoms and molecules, but can here, as with the DNA templates, work with other weaker forces like hydrogen bonds. This reorientation towards hydrogen bonds led to a new branch of chemistry, termed "supramolecular chemistry" by Jean-Marie Lehn in 1978. According to Lehn, the objective of this kind of chemistry is to use hydrogen bonding and stereochemistry to reproduce the selectivity of the interaction between receptors and substrates observed in biology. Thanks to this form of molecular recognition, the building blocks can assemble themselves to form supramolecular structures, and even generate macroscopic materials through these assembly mechanisms.

The Return of Chemistry's Faustian Ambitions

Thus, after having been unceremoniously thrown out the door, nature has made its triumphant return to the chemical laboratory. But has this reconciliation of the chemist with nature also led to the resurrection of the Faustian ambitions formerly associated with both the alchemists and the synthetic chemists?

In contrast to the nineteenth-century chemists who, although capable of synthesizing the substances found in living organisms, were unable to imitate the functioning of nature in their laboratories, today's chemists seem to be well on the way to mastering and reproducing nature's own productive processes. The success of the current intensive research into understanding and copying mechanisms of self-assembly could thus mark a turning point in the longstanding rivalry between chemists and biologists. We need to be clear here, however, that for the modern chemist, mimicking nature does not mean reproducing life. It is no longer a question of proving that life can be reduced to the interplay of chemical forces, thereby shattering the illusion that life is something essentially different from inanimate nature. Chemistry has already won this "victory", at least in the minds and hearts of most scientists. Today, while coming to understand the strategies of synthesis developed during the course of the evolution of life on earth, chemists are using them as models for developing their own "biomimetic" synthetic processes. Thus, the idea of simply copying nature is no longer relevant for contemporary chemists who accept and even emphasize the differences between the strategies used in the evolution of life and those invented in and for the laboratory.

Despite substantial differences in approach, the ambition of today's chemists remains similar to that of their nineteenth-century forbears; they still want to expand their disciplinary territory into that currently held by biology. The boundary between self-assembly and self-organization is easily crossed. In thermodynamic terms, self-assembly is due to the minimization of free energy in a closed system, leading to an equilibrium state. For instance, phospholipids with hydrophobic and hydrophilic ends placed in aqueous solution spontaneously form a stable structure. Self-organization, by contrast, only occurs far from equilibrium and in open systems, as it requires an external energy source. If chemists can manage to control the kinetics of reactions to get complex metastable structures instead of well-ordered materials, they may be able to cross the boundary between chemistry and biology.

The deployment of the whole spectrum of weak forces and the ability to operate at different levels offer other means for chemistry to expand its reach into biology. Indeed, chemists such as Whitesides have become convinced that the most complex phenomena of life can be completely explained by chemistry. The claim is categorical: "The nature of the cells is an entirely molecular problem. It has nothing to do with biology".[18]

The chemists' renewed ambitions rest on a heightened awareness of the collective nature of many chemical properties and a substance's overall behaviour. A glass full of water is different from a single water molecule because isolated molecules do not behave like interacting ones. Jean-Marie Lehn insists on the point that something emerges from these molecules "being together"; a collective behaviour that results from coupling processes rather than being just the expression of the information contained in each individual component. This basic observation has led Lehn to evolve an ambitious project for chemistry, as for him the ultimate aim of this science is to control the basic forces of self-organization. Thus, his program of constitutional dynamical chemistry revives certain ambitions present in the nineteenth century and recalls Berthelot's grandiose program of synthetic chemistry, which was intended to lead him step by step to increasingly complex compounds and, ultimately, to the frontiers of life. Lehn wants to redefine chemistry as the "science of informed matter", a core science mediating inanimate matter (materials processes) and animate matter (living organisms and their complex behaviours).

The Rise of Chemistry's Philosophical Ambitions

Self-assembly deeply transforms the epistemology of chemistry, as chemists no longer adopt the usual positivistic attitude of prudence and are now addressing the big metaphysical questions. As Philip Ball rightly points out, chemists are now talking about the Big Bang and the origins of life.

Far from confining their work to the production of useful materials, chemists want to expand their area of competence to include questions such as the origin of life and even the origin of consciousness. Lehn has indeed suggested that chemistry provides at least a part of the answer to these questions:

> For me, chemistry has a most important contribution to make to the biggest question of all: how does self-organization arise and how does it lead the Universe to generate an entity that is able to reflect on its own origin?[19]

This citation helps us to understand the specificity of the chemists' ambition. First, we should note that during the long history of the disciplinary battle for the title of the "king of the sciences", physicists and chemists have adopted two different strategies. As we saw in Chapter 8, chemists have regularly been led to protest against physicists' claims that they alone can provide the ultimate explanation for natural phenomena. In response to the claim that physics was the fundamental science, chemists launched the counterclaim that theirs was the central science. The justification turns around the ubiquity of chemical phenomena in nature. Because chemistry is everywhere, in living organisms as well as in the non-living environment, it can be seen as a mediator between all the different sectors of scientific knowledge. In 1949, Linus Pauling claimed that "a well-trained chemist — especially a structural chemist — has the best chance of contributing to the integration of the sciences" because, unlike a physicist, he is interested in the different kinds of material structures and can relate individual properties to specific molecular structures.[20] Thus, the most ambitious vision of chemistry is as a science that not only has universal utility, but also serves the unity of science through its ability to federate the sciences, thereby supplanting the physicist's ambition to dominate all the others.

Today, the chemists' approach to natural phenomena — the "knowing through making" approach — seems to be gaining ground among the neighbouring sciences. At the same time as chemists are mimicking biological structures and processes, biologists are starting to mimic chemical techniques. The program of synthetic biology aims at transforming biology in the same way that synthesis transformed chemistry. Reviving the bottom-up conception of synthesis promoted by Berthelot in the nineteenth century, today's synthetic biologists focus on the modelling, design and construction of core components of living systems in order to be able to assemble them into larger integrated systems. The projects are not strictly similar, however, as modern synthetic biologists look at the core components as devices that can be modified or "tuned" to meet specific performance criteria, thereby moving away from the exact imitation of biological products in order to solve technological problems. As an example, we can cite the students from the University of Texas at Austin who engineered a plate of *E-coli* bacteria capable of responding to light to produce a bacterial photography system.

Synthetic biologists have developed an approach in which they break production processes down into their constitutive elements in order to identify the "unit operations" in the synthetic process. In this respect, they seem to have adopted the methods invented by chemical engineers for chemical synthesis a century earlier.[21] Once the pieces of DNA have been redefined in terms of these unit operations, they can then be assembled together to make a module that will behave in a specific way, serving, for instance, as an oscillator or a switch. The overall goal of this approach is to constitute a library of standardized and interchangeable building blocks ("functional biobricks") called the "Registry of Standard Biological Parts". Each one of these units can, in principle, be used to perform a specific function anywhere within a strand of DNA, which can thus cumulate numerous functions. The potential applications of this technique are almost unlimited, although the goal of repairing damaged cells is at the forefront of the uses cited by its promoters. There are a number of scientific visionaries, however, who already see well beyond the modest mission of the medical use of the approach, with champions of synthetic biology like Craig Venter already talking about using these techniques to design living organisms. These artificial organisms would possess a wholly artificial genome made from scratch using DNA synthesis technology. There is also the prospect of a confluence of various nanotechnologies that would allow artificial molecular machines to produce such organisms. Thus, in adopting the program of synthetic chemists, synthetic biologists have reopened the offensive against vitalism, breaking down any remaining barriers between the living and the inanimate.[22] In their effort to make artificial life, they are not just trying to copy biological organisms, they now aim to improve on nature, creating their own novel species. In light of these revolutionary developments in twenty-first century science and technology, is it possible for scientists to resist playing God in this way?

References

1. E. Drexler (1986), p. 13.
2. *Ibid.* p. 5.
3. E. Drexler ed. (1995), p. 2.
4. E. Drexler (1986), p. 14.
5. M. C. Roco, *et al.* (2000).
6. M. Gibbons *et al.* (1994) and H. Nowotny *et al.* (2001).
7. See the articles by Richard Smalley, George Whitesides, and Robert Buderi in *Scientific American,* September 2001.
8. G. Whitesides (2001).
9. R. Jones (2004), pp. 56–86.
10. P. Ball (2002), p. 16.
11. D. E. Clark (ed) (2000).
12. E. Francoeur (2002).
13. X.-D. Xiang *et al.* (1995), and X.-D. Xiang (1999).
14. P. Laszlo (2001), p. 128.
15. J. M. Benyus (1998). P. Ball (2001) and (2002).
16. S. Mann, *et al.* eds (1989), p. 35.
17. M. Amos, 2006.
18. Whitesides quoted by P. Ball (2006), p. 501.
19. Lehn quoted by P. Ball (2006), p. 501.
20. Linus Pauling 'The place of chemistry in the integration of the sciences' (1949) cited in B. Marinacci ed. (1995), pp. 107–111.
21. The concept of unit operation — a basic step in a chemical process — was introduced by Arthur D. Little (1863–1935) a chemical engineer who reformed the training of chemical engineering students at the Massachusetts Institute of Technology. See W. F. Furter ed. (1980).
22. In *Nature* 447, 28 June 2007, the editorial (on pp. 1031–1032) starts with the following statement: 'Synthetic biology provides a welcome antidote to chronic vitalism'.

CHAPTER 14

TOWARDS A RESPONSIBLE
CHEMISTRY

While new technologies may be reviving old alchemical ambitions to penetrate the deepest secrets of life and play God by creating life, we do not want to suggest that all chemists are or have been tainted by this Faustian spirit. One of the background assumptions behind the arguments presented in this book is that there is no eternal essence to chemistry to be traced back across the centuries. The science of chemistry, with its Janus-face image of both modest servant (supplying the materials demanded by modern society) and arrogant creator (a dictatorial technoscience with the ambition of improving on nature) is undoubtedly a cultural production. It has been shaped in specific historical contexts and in turn has deeply influenced our understanding of modernity. Indeed, it is precisely because the values and attitudes promoted by chemists — such as progress through material consumption and unlimited industrial expansion — are associated with the notion of modernity that political movements criticizing modern consumer society regularly target chemistry in their attacks.

Concerns About the Future

A number of recent political tendencies, loosely grouped together under the head of the anti-globalization movement, reject certain aspects of modernity. They claim that modern science and technology (not in themselves, but due to their applications in the context of consumer society and modern capitalism) are responsible for many of the excesses that are leading the world towards political and environmental catastrophe. The dawn of the twenty-first century has been celebrated by a concert of whistle-blowers, including Bill Joy with *The Future Doesn't Need Us*, Martin Rees with *Our Final Hour*, and

Theo Colborn with *Our Stolen Future* to name but three.[1] It is tempting to dismiss these doomsayers as "neo-luddites", the modern-day equivalent of the British textile workers who destroyed the new factory machines in their protest against the changes brought about in their working lives by the industrial revolution. Far from reflecting the conservative views of a movement of disenfranchised artisans, however, these new warnings come from respectable scientists: Bill Joy is a computer scientist, Martin Rees is Britain's Astronomer Royal and Theo Colborn is an expert on endocrine disrupting chemicals. While it is true that there are those who fantasize about self-reproducing nano-machines or robots running amok, these scientists are concerned with the invisible long-term effects of technological advances rather than any single dramatic catastrophic event. Synthetic chemicals are one of the major sources for concern in this context because they have been massively disseminated and consumed in industrial countries over a century, and the long-term effects of this massive distribution of pesticides or solvents might finally become visible.

Just as Rachel Carson denounced the devastating unintended consequences of the indiscriminate use of DDT and other pesticides in the 1950s, Theo Colburn has also recently launched an attack on a variety of chemicals suspected of being endocrine disrupters. Colborn argues that these chemicals, such as bisphenol A, can block the function of various hormones, leading to disease, infertility and even death. Evidence drawn from animals, birds and fish suggests that even trace amounts of such compounds may be responsible for altered sexual behaviour and other problems that can derail the normal process of reproduction. Colburn argues that the persistence of such chemicals in the environment or the dangers associated with the products of their decomposition pose a direct threat to humans as well. Indeed, one of the most audacious arguments proposes such endocrine disrupters as the solution to the enigma of the widespread drop in male fertility across the industrialized world.[2] Nor is Colborn a lone voice; in 2003 dozens of scientists, medical doctors and jurists, ethicists and other citizens, solemnly launched a document known as the "Paris Appeal" (*L'appel de Paris*). Here are the first two articles of their position statement:

Article 1

The development of numerous current diseases is a result of the deterioration of the environment.

Article 2

Chemical pollution represents a serious threat to children and Man's survival.[3]

The authors of the Paris Appeal then went on to call upon decision-makers at all levels to take a number of radical measures such as "banning all products that are certainly or probably thought to be carcinogenic, mutagenic or reprotoxic" for humans, as well as strengthening the regulation applied to industrial chemicals,

The Paris Appeal provoked fierce reactions from chemists who once again felt they were the unfortunate victims of the machinations of other scientists. Chemists rejected the allegations as being unfounded and lacking any base in solid facts. They suggested that the authors of the statement were biased by strong prejudices. Whatever the merits or weaknesses on the two sides, it is clearly no longer acceptable to simply dismiss the calls for the more ethical production and management of chemicals by claiming that modern criticisms are the product of the paranoid fantasies of marginal groups of environmental extremists. Likewise, it is high time to reconsider the philosophy whereby we have no choice but to develop every conceivable scientific and technological advance, whatever the potential risk.

We do have a choice. The widespread desire for an ethics for science — chemistry in particular — in Western societies follows the demise of a blind faith in technological progress that characterized the post-war era. Progress is no longer viewed as a *fatum* or a kind of invisible hand ruling the future of mankind. The future or at least *our* future does need us (all of us). Preserving the future requires that individual citizens take an active interest and exercise their enlightened judgement concerning decisions that need to be made today. It is no longer sufficient to rely exclusively on the advice of a handful of experts in risk assessment.

The Chemist's Code of Conduct

Significantly, one of the most powerful professional organizations of chemists, the American Chemical Society (ACS), which had proclaimed its faith in progress in *The Chemist's Creed* published in 1965, adopted a *Chemist's Code of Conduct* for the guidance of its members in 1994.

Chemists Acknowledge Responsibilities To:

The Public
Chemists have a professional responsibility to serve the public interest and welfare and to further knowledge of science. Chemists should actively be concerned with the health and welfare of co-workers, consumers and the community. Public comments on scientific matters should be made with care and precision, without unsubstantiated, exaggerated, or premature statements.

The Science of Chemistry
Chemists should seek to advance chemical science, understand the limitations of their knowledge, and respect the truth. Chemists should ensure that their scientific contributions, and those of their collaborators, are thorough, accurate, and unbiased in design, implementation, and presentation.

The Profession
Chemists should remain current with developments in their field, share ideas and information, keep accurate and complete laboratory records, maintain integrity in all conduct and publications, and give due credit to the contributions of others. Conflicts of interest and scientific misconduct, such as fabrication, falsification, and plagiarism, are incompatible with this code.

The Employer
Chemists should promote and protect the legitimate interests of their employers, perform work honestly and competently, fulfill obligations, and safeguard proprietary information.

Employees
Chemists, as employers, should treat subordinates with respect for their professionalism and concern for their well-being, and provide them with a safe, congenial working environment, fair compensation, and proper acknowledgement of their scientific contributions.

Students
Chemists should regard the tutelage of students as a trust conferred by the society for the promotion of the student's learning and professional

development. Each student should be treated respectfully and without exploitation.

Associates
Chemists should treat associates with respect, regardless of the level of their formal education, encourage them, learn with them, share ideas honestly, and give credit for their contributions.

Clients
Chemists should serve clients faithfully and incorruptibly, respect confidentiality, advise honestly, and charge fairly.

The Environment
Chemists should understand and anticipate the environmental consequences of their work. Chemists have the responsibility to avoid pollution and to protect the environment.[4]

This charter, adopted by the board of directors of the ACS, provides clues about how today's chemists view their own mission and allegiances. Interestingly, they place concern with the public good first, even before the advancement of chemical knowledge. Just like the eighteenth-century pioneers of chemical science, chemists of today think that they have the ability and a responsibility to advance the cause of public welfare. In this sense, chemistry remains a civic science, as chemists assume a responsibility for the comfort and health of society at large.

While professional duties are a secondary priority, they nevertheless constitute the most substantial part of the charter with a long list of allegiances to the profession, employers and employees, students (future professionals), associates and clients. The chemist's professional ethos encompasses most of the usual prescriptions for good conduct and prohibitions of misconduct that are generally associated with the scientific ethos. However, the charter does not explain what to do should such rules for good conduct in science enter into conflict with demands stemming from one's allegiance to the employer or to other employees. Indeed, in chemistry, as in other applied sciences, theoretical or pure research and industrial interests are often intertwined, and researchers are regularly confronted by conflicts of interest. The four basic principles of the scientist's ethos as proposed by the sociologist Robert Merton in 1942 — universalism, communism, disinterestedness and organized scepticism — are ideals that are rarely realized on the ground.[5] A significant amount of research results are neither circulated nor published for reasons of industrial secrecy, and scientific expertise is often biased by private interests. Chemical research is rarely disinterested as it is often driven by the hope of making money from patents or via other types of intellectual property.

The last and least important concern addressed in the American Chemical Society's code of conduct is with the environment. Clearly, this issue is not a priority. Chemists are, however, obliged to integrate environmental issues into their research agenda because of the legal and societal pressures that have been mounting over recent decades. Indeed, since the 1950s, government regulatory bodies have become increasingly concerned about the way in which brand new and untested chemicals are introduced onto the market. Recently, the European Union has felt it necessary to launch a bold new initiative to curb the indiscriminate production and dissemination of new synthetic chemicals: the REACH program.

From Prudence to Precaution

The REACH regulation (REACH is an acronym for Registration, Evaluation, Authorisation and Restriction of Chemicals) has been in force since June 2007, and requires all companies manufacturing or importing more than one metric tonne of a chemical substance per year to register itself in a central database administered by the new European Union Chemicals Agency.[6] The registrants must also identify appropriate risk management measures and communicate them to the users. These measures were the result of a wide-ranging consultation open to all the stakeholders. As a result, they represent a compromise between the interests of the chemical companies, mainly concerned about their competitiveness in an increasingly global market, and the social concern for protecting human health and the environment. Thus, the REACH regulation can be seen as a reminder that control and risk-management are central concerns in chemistry because all chemical substances are potentially hazardous. Registration and risk assessment is intended to allow government and non-government agencies to trace and control chemicals in the environment, with the idea that chemical companies can use this information to ensure they respect a margin of safety, and limit the production of chemicals to keep the concentrations at a level at which no adverse effect have been observed or predicted. REACH also advocates a high level of transparency, demanding the passage of all relevant safety information along the whole chain of supply from the manufacturer to the end-user or consumer. This, however, is just a part of the regulation, and the part that bothers the chemical industry the least. The REACH program goes further and stipulates that supplementary evaluations need to be made for familiar substances that could potentially cause cancer, infertility or birth defects, as well as those that do not degrade and so accumulate in the environment. More importantly, the REACH program transfers the responsibility for this kind of research from government agencies to the chemical companies themselves.

It should be clear, therefore, that the REACH program instantiates the attitude of precaution that prevails in European politics around questions concerning public health and the environment. This sense of precaution has to be understood as a contrast to more conventional attitudes such as foresight and prevention. Insurance companies specialise in foresight, as they fix their clients' payments based on anticipated future pay-outs for the accidents that cannot be predicted individually but which they know will keep on happening. Prevention is the daily business of engineers and manufacturers, and so they weigh up the costs and benefits involved in each process or technique under consideration. Risk assessment concerns situations where it is possible to display the list of possible scenarios and calculate the potential harm and benefit associated with each one. In such a case, the rational choice can be based on the application of probability calculus, and the choice to avoid an unacceptably high risk, however improbable, can be seen as elementary prudence. By contrast, the principle of precaution concerns situations where probability calculus is not applicable.[7] This is the case when we know that we do not have the complete list of potential scenarios, or when we know that we have not identified all the factors and parameters that might determine a catastrophe. In these circumstances, we do not have the necessary elements for assessing the risk and have to make a decision and take action in the absence of the scientific evidence we might desire. In short, precaution is considered the appropriate response to radical uncertainty when the stakes are very high.

This ethical attitude has been raised up to the status of a political principle. The precautionary principle underlying the Rio de Janeiro Declaration on Environment and Development (1992) and included in the European Maastricht Treaty on the Environment signed in 1992, states that if an action or policy might cause severe or irreversible harm to the public, and in the absence of a scientific consensus that harm would not ensue, the burden of proof falls on those who would advocate taking the action. In fact, there is no consensus about the terms of the principle, and a wide range of interpretations are currently used. The weakest interpretation of the principle reduces it to an application of a cost-benefit analysis, whereas the strictest one prohibits any action that presents any potential for causing significant harm, unless the proponent of the action can demonstrate that it presents no appreciable risk of harm.

The principle of precaution is mainly invoked in cases where one has to decide whether or not to introduce an innovation with potentially severe and irreversible negative consequences, such as the use of chemicals in water treatment or the introduction of genetically modified organisms into the environment. Evidently, chemistry is one of the major areas in which such cases have been identified because of the long-term impact of chemicals on the biosphere. Furthermore, synthetic chemists have always lived, and always will live in a world of uncertainties because the end products of their creative minds belong at the same time to the complex ecosystem of nature and to a social environment that is out of their control and remains largely unpredictable. Although the proponents of the principle often have to face the accusation of being anti-science, chemists have gradually come to realize that they can no longer be satisfied with their "prudent" approach to managing chemicals, and they are going to be obliged to adopt a precautionary attitude.

A New Chemical Culture?

One immediate and common response to the question of how to render an industry that pollutes more responsible is to make it pay for the negative consequences of such pollution. Ethylene, for example, is a vital chemical used in the manufacture of a variety of products. To obtain large quantities of ethylene, companies use tonnes of oil — a non-renewable natural product that produces carbon dioxide when burned — and run huge steam-crackers that are extremely polluting. Thus, according to this logic of making the polluter pay, ethylene production should be taxed, and the money raised should be used to fund relevant research or in decontamination projects. This, however, represents a lazy attitude that does not encourage a search for any less polluting process and precludes all questions about the appropriateness of a technological system based on oil, a system in which ethylene constitutes a major and critical ingredient. The political questions of how we should live and what should or should not be manufactured are thereby evacuated from the debate.

An alternative answer would be to proscribe the causes that threaten human and environmental health, thereby extending the Hippocratic principle of medical ethics *primum non nocere* (first, do not harm) to professional chemists. In their efforts to treat their patients, medical doctors should, according to this dictum, ensure that they do not harm them even if they are not able to cure them. Similarly, in their noble effort to improve on nature, chemists should, at the very least, not degrade the environment they are attempting to ameliorate.

There is another way to address this question of the chemists' responsibility, however, by adopting an approach that avoids generating environmental hazards in the first place. This approach asks chemists to think about the long term consequences of their actions and productions even before they introduce new manufacturing processes or develop new synthetic products. This attitude demands that chemists go beyond the conventional claims of morality based on their good intentions. Of course chemists are motivated by a concern for public health and welfare, who isn't? Nevertheless, one does not have to look very far to see that good will is not enough. Responsible chemists have to consider and anticipate the adverse, long-term and unintended consequences of their actions.

Let us take the example of Freon, which used to be widely used in refrigerators. This synthetic chemical was selected for this use because of its properties that made it a good fluid for operating the heat exchange mechanisms involved in refrigeration. This substance has proved to be an environmental hazard as it reacts with the ozone in the atmosphere, triggering a chain reaction transforming ozone into oxygen. The resulting "holes" in the ozone layer were identified as a threat to human health, and Freon, among other chemicals, was banned. As a result, chemists developed alternative products that have replaced Freon in refrigerators but do not degrade the ozone layer when released into the atmosphere. Here, we have a classic story of a technological response to an environmental problem. But it is a story of a reaction to a problem, and does not involve sufficient proactive investigation and reasoning aimed at avoiding the problem in the first place. What we are advocating as a new ethical approach in chemistry — one that is already being adopted by many public and private laboratories — is to actively identify the potential problems from the inception of the chemical synthesis. While it is impossible to predict all the possible negative consequences of a chemical reaction or product, there needs to be more effort to consider all the known problems. There are plenty of pollutants that have already been identified; mercury, dioxins, and, perhaps the most important in recent times, carbon dioxide and other greenhouse gases. It is evident that chemical processes need to be developed with the aim of not introducing these substances into the environment. The production of substitutes for polluting materials or alternatives to polluting processes is one aspect of this approach, but another, more important aspect is the encouragement of innovative research that takes into account the possible consequences of the process or product from the very beginning. Thus, chemists should not follow their traditional strategy of resisting bans on dangerous substances, as was the case with Freon and other CFCs, but should embrace the identification of such environmental hazards and make every possible effort to replace them by alternatives. Furthermore, research-based chemical companies should not resist or cover up the identification of such threats, but rather be at the forefront of this work and integrate this research directly into their strategies of product development. The aim of producing non-harmful products should be at the top of their priorities and not just another cost-benefit factor. The future of the chemical industry and perhaps of humanity itself depends on this kind of shift in industrial strategy.

In this context, it is vital to consider the whole life of chemicals, in particular what effects they might have when released into the environment after use. These kinds of ethical consideration do not, however, exclusively concern industrial chemists. In basic research, chemists are often driven by a passion for creating new molecules or for meeting big synthetic challenges and do not think enough about the consequences of their research. They need to pose the difficult question: what will become of these non-natural molecules once they have been created? Even confined to a laboratory or stored in a library of molecules, they still exist, and the chemists are responsible for people who might use them not only in this generation, but also in future generations, as well as the various unpredictable uses that can be found in times of political strife or war. This is even more the case when chemists are creating hybrid artefacts using organic parts such as bacteria proteins and micelles to perform specific tasks, since they can have unpredictable interactions with living tissues or living organisms in the environment.

Finally, a responsible chemist should care about ethical values. Ethics is, after all, about the just or the "good" life; this simple observation allows us to reinterpret DuPont's famous slogan "Better things for better living … through chemistry" from an alternative perspective that does not lead to the unreflective mass production and mass consumption of synthetic chemicals. In order to participate actively in a better life, responsible chemists need to pay close attention to the human values implicit in their material (and theoretical) productions.

In fact, one of the major reasons for the rise of the chemical industry has been the chemists' ability to anticipate and respond to (as well as sometimes to construct) the societal expectations for what counts as the good life. Over the past century, chemists have promoted such values as improvement and progress against an ideological backdrop in which material production and consumption served as the principal indicators of civilisation. Will chemists be able to transform the values attached to their activity in order to adapt to new societal demands for the protection of the environment and the preservation of life on earth? These goals require a different attitude with regard to natural resources. Instead of using molecules as tools, devices or machines that work for the principal if not exclusive benefit of humans, the material resources to be found in the periodic system have to be considered as genuine partners engaged in a shared project. Chemists are bound to them by a kind of contract, which entitles these molecules (both existing and potential) to care and respect in their own right. Chemists can no longer unreflectively take advantage of the sublime properties of molecules. It is not enough to try and clean up polluted sites, respect for the environment requires anticipating risks (to human and environmental health) at the stage of designing new processes or products. From this perspective, chemicals would no longer be considered as the antonym of "natural", and chemists would no longer be seen as the enemies of the environmental movement. Instead, chemistry could be considered to be a branch of the public service in charge of taking care of natural resources.

Chemists should not, therefore, be satisfied with decontaminating polluted sites and claiming that chemistry does not harm the environment. Furthermore, when considering the impact of nanochemistry, it is insufficient to limit one's analysis to the conclusion that this technology will reduce the quantity of raw materials consumed by industrialized countries. It is not unreasonable to suspect that the process of dematerialization associated with nanotechnology may be nothing more than science-fiction or, worse still, a simple repackaging of the old dualistic paradigm that proclaims the domination of mind over matter. In this vision, with which we are only too familiar, new materials and techniques are just so many more tools in the hands of the chemist-demiurge with which he can remake the world as society wants it. Now, however, chemists are in the situation of being able to engage material individuals as partners in their cognitive and technological enterprises. Thus, they find themselves in the novel position of being able to envisage molecules as natural beings with which they enter into a partnership rather than as mere devices or tools in the service of men.

Furthermore, if chemistry is to have a chance of being considered as a science which is in the service of the public, it seems reasonable for chemists and chemical industries to share their responsibility with citizens. In practical terms, this means that scientific and technological choices should no longer be left to chemists alone, and should rather be shared as collective decisions. The good governance of chemicals demands the engagement of the public in science and technology policy, allowing the development of new sociopolitical forums in which the impurity of chemistry could be turned into a positive feature. The entangled mixture of nature and artefact, nature and society, as well as science and technology that inspired the image of chemistry as an imperfect and impure science in the twentieth century could make it the model science for the twenty-first century. It is, then, possible to envisage a new configuration of chemistry reshaped as a technoscience capable of integrating culture and society into its practice, and thereby able to overcome the secular conflict between nature and artefact.

References

1. B. Joy (2000), M. Rees (2003), T. Colborn *et al.* (1996).
2. For the arguments around endocrine disrupters, see T. Colborn *et al.* (1996), and for an analysis of the debate that has arisen around this question and how it has been conducted, see S. Krimsky (2000).
3. The Paris Appeal 'http://artac.info' (accessed on September 17, 2007).
4. 'www.portal.acs.org' (accessed on February 22, 2007). A slightly modified version, 'The Chemical Professional's Code of Conduct', was adopted in March 2007.
5. R. K. Merton (1942).
6. 'http://ec.europa.eu/enterprise/reach/index_en.htm', (accessed on September 17, 2007).
7. See M. Callon *et al.* (2001).

BIBLIOGRAPHY

Adam, David (2001) 'What's in a Name?' *Nature*, **411**, May 24: 408–409.

Aftalion, Fred (1991) *A History of the International Chemical Industry*, trans. O. T. Benfey, Philadelphia, University of Pennsylvania Press.

Amos, Martyn (2006) *Genesis Machines: The New Science of Biocomputing*, Atlantic Books.

Andersen, A. (1998) 'Pollution and the chemistry industry: The case of the German dye industry' in E. Homburg, A. Travis and H. G. Schröter (eds.) (1998) pp. 183–200.

Andrade Martins, Roberto de (1993) 'Os experimentais de Landolt sobre a conservação da massa' *Quimica Nova* **16**: 481–490.

Arendt, Hannah (1958) *The Human Condition*, Chicago, The University of Chicago Press.

Aristotle (350 BCE) *Physics II*, Translated by R. P. Hardie and R. K. Gaye, available in electronic form on http://classics.mit.edu/

Aristotle (n.d.) *De Generatione et Corruptione* available in English translation on http://classics.mit.edu/

Bachelard, Gaston (1930) *Le pluralisme cohérent de la chimie moderne*, reprint Paris, Vrin, 1973.

Bachelard, Gaston (1938) *La formation de l'esprit scientifique*, reprint Paris, Vrin, 1971.

Bachelard, Gaston (1953) *Le matérialisme rationnel*, reprint Paris, PUF, 1990.

Bacon, Francis (1620) *Novum Organum*, reprint New York, Colonial Press, 1899.

Baird, Davis (1993) 'Analytical Chemistry and the Big Scientific Instrumentation Revolution' *Annals of Science*, **50**: 267–290.

Baird, Davis, Eric Scerri, and Lee McIntyre (eds.) (2006) *Philosophy of Chemistry. Synthesis of a New Discipline,* Dordrecht, Springer.

Ball, Philip (2001) *The Self-Made Tapestry: Pattern Formation in Nature, Oxford,* Oxford University Press.

Ball, Philip (2002) 'Natural strategies for the molecular engineer' *Nanotechnology*, **13**, 15–28.

Ball, Philip (2006) 'What chemists want to know' *Nature*, **442/3**, August 2006: 500–502.

Barthes, Roland (1971) *Mythologies*, Paris, Denoel-Gonthier, Transl. London, Vintage, 1993.

Baud, P. (1932) *L'industrie chimique en France*, Paris, Masson & Cie.

Baudrillard, Jean (1968) *Le système des objets*, Paris, Gallimard, reprint 2000.

Beck, Ulrich (1992) *Risk Society: Towards a New Modernity*, London, Sage.

Béguin, Jean (1610) *Élements de chymie*, Paris.

Bensaude-Vincent, Bernadette (1982) 'L'éther, élément chimique: un essai malheureux de Mendeleev en 1904' *British Journal for the History of Science*, **15**: 183–187.

Bensaude-Vincent, B. (1986) 'Mendeleev's periodic system of chemical elements' *British Journal for the History of Science*, **19**: 3–17.

Bensaude-Vincent, B. (1992) 'The balance: Between chemistry and politics' *Eighteenth Century*, **33**, No. 3: 1992, 217–237.

Bensaude-Vincent, B. (1993) *Lavoisier, Mémoires d'une révolution*, Paris, Flammarion.

Bensaude-Vincent, B. (1994) 'La chimie, un statut toujours problématique dans la classification du savoir' *Revue de Synthèse,* **115**: 135–148.

Bensaude-Vincent, B. (1996) 'Between history and memory: Centennial and bicentennial images of Lavoisier' *Isis*, **87**: 481–499.

Bensaude-Vincent, B. (1998) *Eloge du mixte. Matériaux nouveaux et philosophie ancienne*, Paris, Hachette Littératures.

Bensaude-Vincent, B. (1999) 'Atomism and Positivism: A legend about French Chemistry' *Annals of Science*, **56**: 81–94.

Bensaude-Vincent, B. (2003) 'A Language to order the chaos' in M. Jo Nye (ed.), *The Cambridge History of Science, Vol V. Modern Physical and Mathematical Sciences*, Cambridge, Cambridge University Press, 174–190.

Bensaude-Vincent, B. and Ferdinando Abbri (eds.) (1995) *Lavoisier in European Context. Negotiating a New Language for Chemistry*, Cambridge, Science History Publications.

Bensaude-Vincent, B. and Isabelle Stengers (1996) *A History of Chemistry*, Cambridge, MA, Harvard University Press.

Bensaude-Vincent, B., H. Arribart, Y. Bouligand and C. Sanchez (2002) 'Chemists at the School of Nature' *New Journal of Chemistry*, **26**: 1–5.

Bensaude-Vincent, B. and Bruno Bernardi (eds.) (2003) *Rousseau et les sciences*, Paris, L'Harmattan.

Benyus, Janine M. (1998) *Biomimicry, Innovation Inspired by Nature*, New York, Quill edition.

Berthelot, Marcellin (1876) *La synthèse chimique*, Paris, Alcan.

Berthelot, Marcellin (1877) 'Réponse à la note de M. Wurtz, relative à la loi d'Avogadro et à la théorie atomique' *Comptes-rendus de l'Académie des sciences*, **84**: 1189–1195.

Black, Joseph (1754) *De Humore Acido a Cibis orto, et Magnesia Alba*, Edinburgh, MD thesis.

Blondel-Mégrelis, Marika (1996). *Dire les choses. Auguste Laurent et la méthode chimique*, Paris, Vrin.

Boas-Hall, Marie (1965) *Robert Boyle on natural Philosophy*, Bloomington IN, Indiana University Press.

Boas-Hall, Marie (1968) 'The history of the concept of element' in D. L. S. Cardwell (ed.), *John Dalton and the Progress of Science*, Manchester, Manchester University Press, pp. 21–39.

Boerhaave, Hermann (1745) *Elementa chemiae*, Leyden, French transl. *Elémens de chymie*, Paris, 1748.

Boltanski, Luc, and Eve Chiapello (2000) *Le nouvel esprit du capitalisme*, Paris, Gallimard.

Bougard, Michel (1999) *La chimie de Nicolas Lemery, apothicaire et médecin (1645–1715)*, Bruxelles, Brepols.

Boulding, K. E. (1966) 'The Economics of the Coming Spaceship Earth' in *Environment Quality in a Growing Economy*, Baltimore, Johns Hopkins University Press.

Bourdieu, Pierre (1979) *Distinction: a social critique of the judgment of taste*, London, Routledge and Kegan Paul.

Boyle, Robert (1661) *The Sceptical Chymist*, London, Cadwell and Crooke.

Brenner, Anastasios (2003) *Les origines françaises de la philosophie des sciences*, Paris, PUF.

Brickman, R, S. Jasanoff, and T. Ilgen (1985) *Controlling Chemicals: The politics of Regulation in Europe and the United States*, Ithaca, Cornell University Press.

Brock, William H. (1997) *Justus von Liebig: The Chemical Gatekeeper*, Cambridge, Cambridge University Press.

Brooke, John Hedley (1968) 'Wöhler's Urea and its Vital Force — A Verdict from the Chemists.' *Ambix*, **15**: 84–114.

Brooke, John Hedley (1995) *Thinking about Matter*, Aldershot, Ashgate Variorum.

Buès, Christiane (2000) 'Histoire du concept de mole (1869–1969) à la croisée des disciplines physique et chimie' *L'Actualité chimique*, October 2000: 39–42.

Bushan, Nalini, and Stuart Rosenfeld (eds.) (2000) *Of Minds and Molecules: New Philosophical Perspectives on Chemistry*, Oxford, Oxford University Press.

Butterfield, Herbert (1957) *Origins of Modern Science: 1300–1800*, New York, Macmillan.

Callon, Michel, P. Lascousmes, and Y. Barthes (2001) *Agir dans un monde incertain*, Paris, Seuil.

Cardwell, D. S. L. (1975) 'Science and World War I' *Proceedings of the Royal Society of London*, **A**, **342**: 447–456.

Carneiro, Ana (1993) 'Adolphe Wurtz and the atomism controversy' *Ambix*, **40**: 75–93.

Carson, Rachel (1962) *Silent Spring*, Boston, Houghton Mifflin.

Cartwright, Nancy (1983) *How the Laws of Physics Lie*, Oxford, Oxford University Press.

Cartwright, Nancy (1989) *Nature's Capacities and their Measurement*, Oxford, Oxford University Press.

Cassirer, Ernst (1953) *Substance and Function and Einstein's Theory of relativity*, New York, Dover Publications. Published in French as *Substance et fonction. Éléments pour une théorie du concept*, transl. Pierre Caussat, Paris, Minuit, 1910.

Champetier, George (1940) 'L'évolution de la chimie' *Les cahiers rationalistes*, **8**: 5–30.

Chaptal, Jean-Antoine (1807) *Chimie appliquée aux arts*, Paris, Déterville.

Clark, David E. (ed.) (2000) *Evolutionary Algorithms in Molecular Design*, Weinheim, Wiley-VCH.

Clave, Etienne de (1641) *Nouvelle lumière philosophique*, reprint Paris, Fayard, 1999.

Clericuzio, Antonio (1993) 'From Van Helmont to Boyle: A Study of the Transmission of Helmontian Chemistry and Medical Theories in Seventeenth-Century England' *British Journal for the History of Science*, **26**: 303–334.

Clericuzio, Antonio (2000) *Elements, Principles and Corpuscles. A Study of Atomism and Chemistry in the Seventeenth Century*, Dordrecht, Boston, Kluwer.

Clow, Archibald and Nan L. Clow (1952) *The Chemical Revolution: A Contribution to Social Technology*, reprint London, Gordon and Breach, 1992.

Cognard, Philippe (1989) *Les applications industrielles des matériaux composites*, Paris, Editions du Moniteur, 2 vols.

Cohendet, Patrick, *et al.* (1984) *La chimie en Europe, innovations mutations et perspectives*, Paris, Economica.

Colborn, Theo, Diane Dumanoski, and John Peter Myers (1996) *Our Stolen Future: Are We Threatening Our Fertility, Intelligence and Survival-A Scientific Detective Story*, New York, Penguin Books.

Collins, Harry (1985) *Changing Order: Replication and Induction in Scientific Practice*, Beverley Hills & London, Sage.

Comte, Auguste (1830–1842) *Cours de philosophie positive*, Paris, 6 vols. reprint Paris, Hermann, 1975, 2 vols.

Comte, Auguste (1844) *Discours sur l'esprit positif*, Paris, reprint Paris, Vrin, 1995.

Condillac, Etienne Bonnot de (1780) *La Logique*, reprinted in Raymond Bayer ed. *Œuvres philosophiques de Condillac*, Paris, PUF, Vol. 2, 1948.

Crosland, Maurice (1967) *The Society of Arcueil: A View of French Science at the Time of Napoleon*, Cambridge MA, Harvard University Press.

Crosland, Maurice (ed.) (1971) *The Science of Matter. Selected Readings*, reprint London, Gordon & Breach, 1992.

Dagognet, François (1969) *Tableaux et langages de la chimie*, Paris, Vrin.

Dagognet, François (1985) *Rematérialiser. Matières et matérialismes*, Paris, Vrin.

Darnton, Robert (1979) *The Business of Enlightenment: A Publishing History of the Encyclopédie, 1775–1800*, Cambridge MA, Harvard University Press.

Daston, Lorraine and Katharine Park (1998) *Wonders and the Order of Nature 1150–1750*, New York, Zone Books.

Daumas, Maurice (1946) *L'acte chimique. Essai sur l'histoire de la philosophie chimique*, Bruxelles, éditions du Sablon.

Daumas, Maurice (1955) *Lavoisier, théoricien et expérimentateur*, Paris, PUF.

Daumas, Maurice and D. I. Duveen (1959) 'Lavoiser's relatively unknown large scale experiment of decomposition and synthesis of water. February 27–28, 1785' *Chymia*, **5**, 111–129.

Debus, Allen G. (1967) 'Fire analysis and the elements in the sixteenth and seventeenth centuries' *Annals of Science, 23*: 127–147.

Debus, Allen G. (2006) 'Chemical Medicine in Early Modern Europe' in *The Chemical Promise: Experiment and Mysticism in the Chemical Philosophy*, Sagamore Beach, Science History Publications, pp. 63–97.

Delacre, Maurice (1923) *Essai de philosophie chimique*, Paris, Payot.

Descartes, René (1647) *Principes de la philosophie, in* Adam-Tannery (eds.) *Oeuvres*, Paris, Vrin, Vol. 9, 1964.

Descartes, René (1628–1629) *Règles pour la direction de l'esprit*, transl. J. Sirven, Paris, Vrin, 1990.

Diderot, Denis (1753) *Discours sur l'interprétation de la nature*, in *Œuvres philosophiques*, Paris, Garnier, 1964.

Diderot, Denis, and Jean D'Alembert (eds.) (1751–1765) *Encyclopédie ou dictionnaire raisonné des sciences des arts et des métiers par une société de gens de lettres*, Paris, Panckoucke.

Dirac, P. A. M. (1929) 'Quantum mechanics of many-electron systems' *Proceedings of the Royal Society of London* **A**, **123**: 714–33.

Donovan, Arthur (1993) *Antoine Lavoisier: Science, Administration and Revolution*, Oxford, Blackwell.

Drexler, Eric (1986) *Engines of Creation*, New York, Anchor Books, 2nd edition, 1990.

Drexler, Eric (1995) 'Introduction to nanotechnology' in Markus Krummenacker and James Lewis (eds.), *Prospects in Nanotechnology. Proceedings of the 1st general conference on nanotechnology: Developments, applications, and opportunities, Palo-Alto, 1992,* New York, John Wiley & Sons.

Duhem, Pierre (1892) 'Notation atomique et hypothèses atomistiques' *Revue des questions scientifiques*, **31**: 391–454.

Duhem, Pierre (1902) *Le mixte et la combinaison chimique*, English translation in P. Needham *Mixture and Chemical Combination and Related Essays*, Dordrecht, Kluwer, 2002

Duhem, Pierre (1906) *La théorie physique, son objet, sa structure*, Transl. *The Aim and Structure of Physical Theory*, Princeton, Princeton University Press, 1954.

Duhem, Pierre (1916) *La chimie est-elle une science française?* Paris, Hermann.

Dumas, Jean-Baptiste (1837) *Leçons sur la philosophie chimique*, Paris; reprint, Brussels, Culture et civilisation, 1972.

Duncan, Alistair W. (1988) 'Particles and Eighteenth-Century Concepts of Chemical Combination' *British Journal for the History of Science*, **21**: 447–453.

Duncan, Alistair W. (1996) *Laws and Order in Eighteenth-Century Chemistry*, Oxford, Clarendon Press.

Emptoz, Gérard, and Patricia Aceves Pastrana (eds.) (2000) *Between the Natural and the Artificial. Dyestuffs and Medicine*, Bruxelles, Brepols.

Fajans, Kasimir (1913) 'Radioactive Transformations and the Periodic System of The Elements' *Berichte der Deutschen Chemischen Gesellschaft*, **46**: 422–439.

Farrar, W. V. (1965) 'Nineteenth-Century speculations on the complexity of chemical elements' *British Journal for the History of Science*, **2**: 297–323.

Feyerabend, Paul (1965) 'On the meaning of scientific terms' *Journal of Philosophy*, **62**: 266–274.

Fjors, Hjalmar (2003) *Mutual Favours: The Social and Scientific Practice of Eighteenth-Century Swedish Chemistry*. PhD. Diss. Uppsala University.

Fontenelle, Bernard Le Bovier de (1686) *Entretiens sur la pluralité des mondes habités*, reprint Paris, Fayard, 1991.

Foucault, Michel (1977) *Discipline and Punish: the Birth of the Prison*, Trans. Alan Sheridan, New York, Pantheon Books.

Francoeur, Eric (2002) 'Cyrus Leventhal, the Kluge and the origin of interactive molecular graphics' *Endeavour*, **26**, No. 1: 127–131.

Friedel, Robert (1983) *Pioneer Plastic: The Making and Selling of Celluloid*, Madison, University of Wisconsin Press.

Furter, William F. (ed.) (1980) *History of Chemical Engineering*, Washington DC, American Chemical Society.

Fustel de Coulanges, Numa (1888) *Histoire des institutions de l'ancienne France*, reprint, Brussels, Culture et civilisation, 1964.

Galison, Peter, and D. J. Stump (eds.) (1996) *Disunity of Science, Boundaries, Contexts and Power*, Stanford, Stanford University Press.

Gavroglu, Kostas, and Ana Simoes (1994) 'The Americans, the Germans, and the Beginnings of Quantum Chemistry: The Confluence of Diverging Traditions' *Historical Studies in the Physical Sciences*, **25**: 47–110.

Gibbons M., C. Limoges, H. Nowotny, *et al.* (1994) *The New Production of Knowledge*, London, Sage Publications, 2nd ed. 1996.

Ginzburg, Carlo (1989) 'Traces. Racines d'un paradigme indiciaire' in *Mythes, emblèmes, traces*, Paris, Flammarion, pp. 139–180.

Golinski, Jan (1992) *Science as Public Culture. Chemistry and Enlightenment in Britain, 1760–1820*, Cambridge, Cambridge University Press.

Görs, Britta (1999) *Chemischer Atomimsus: Anwendung, Veränderung, Alternativen in deutschsprachigen Raum in der zweiten Hälfte des 19 Jahrhunderts*, Berlin.

Goupil, M. (1991) *Du flou au clair? Histoire de l'affinité chimique*, Paris, Éditions du CTHS.

Grapi, Pere (2001) 'The Marginalization of Berthollet's chemical affinities in the French textbook tradition at the beginning of the nineteenth century' *Annals of Science*, **58**, 115–136.

Gras, Alain (2003) *La fragilité de la puissance. Se libérer de l'emprise technologique*, Paris, Fayard.

Guyton de Morveau, Louis-Bernard, Antoine-Laurent Lavoisier, Claude-Louis Berthollet and Antoine-François de Fourcroy (1787) *Méthode de nomenclature chimique*, Cuchet, Paris, reprint, Paris, Seuil, 1994.

Haber, Ludwig F. (1958) *The Chemical Industry during the Nineteenth-Century*, Oxford, Clarendon Press.

Haber, Ludwig F. (1986) *The Poisonous Cloud. Chemical Warfare in the First World War*, Oxford, Clarendon Press.

Hacking, Ian (1983) *Representing and Intervening*, Cambridge, Cambridge University Press.

Hamlin, Christopher (1993) 'Between knowledge and action: Themes in the history of environmental chemistry' in Seymour H. Mauskopf (ed.), *Chemical Sciences in the Modern World*, Philadelphia, University of Pennsylvania Press, pp. 295–321.

Hannaway, Owen (1975) *The Chemist and the Word: The Didactic Origins of Chemistry,* Baltimore, Johns Hopkins University.

Handley, Susannah (1999) *Nylon The Story of a Fashion Revolution*, Baltimore, The Johns Hopkins University Press.

Haynes, Roslynn D. (1994) *From Faust to Strangelove: Representations of the Scientists in Western Literature*, Baltimore, The John Hopkins University Press.

Heidegger, Martin (1954) 'The Question Concerning Technology' in William Lovitt (ed.), The *Question Concerning Technology and Other Essays*, New York, Harper, 1977, pp. 3–35.

Hiebert Erwin (1971) 'The energetics controversy in late nineteenth century Germany' in D. H. Roller (ed.) *Perspectives in the History of Science and Technology*, Norman, University of Oklahoma Press, pp. 67–86.

Hoddeson L., E. Braun, J. Teichman, and S. Weart (eds.) (1992) *Out of the Crystal Maze. Chapters from the History of Solid State Physics,* Oxford, Oxford University Press.

Hoffmann, Roald (1995) *The Same and not the Same*, New York, Columbia University Press.

Hoffmann, Roald (2001) 'Not a library' *Angewandte Chemie, International Edition,* **40**, No. 18: 3337–3340.

Holmes, Frederic L. (1962) 'From elective affinity to chemical equilibrium: Berthollet's laws of mass action' *Chymia*, **8**: 105–145.

Holmes, F. L. (1971) 'Analysis by Fire and Solvent Extractions: The Metamorphosis of a Tradition' *Isis* **62**: 129–148.

Holmes, F. L. (1989) *Eighteenth-century Chemistry as an Investigative Enterprise,* Berkeley, Office for the History of Science and Technology, University of California.

Holmes, F. L. (1995) 'Concepts, operations and the problem of 'Modernity' in early modern chemistry'. Paper presented at the workshop on early modern chemistry, Max Planck Institute, Berlin, June 1995.

Holmes, F. L. (1996) 'The communal context for Etienne-François Geoffroy's *Table des rapports*' *Science in Context,* **9**: 289–311.

Holmes, Frederic L. and Trevor H. Levere (eds.) (2000) *Instruments and Experimentation in the History of Chemistry,* Cambridge MA, The MIT Press.

Homberg, Wilhelm (1703) 'Essai sur l'analyse du souffre commun' in *Mémoires de l'Académie royale des sciences de Paris,* pp. 31–40.

Homburg, Ernst, Anthony Travis and Harm G. Schröter (eds.) (1998) *The Chemical Industry in Europe, 1850–1914, Industrial Growth, Pollution and Professionalization,* Dordrecht, Kluwer Academic Publishers.

Hooykaas, Robert (1972) *Religion and the Rise of Modern Science,* Edinburgh, Scottish Academic Press.

Jacques, Jean (1954) 'La thèse de doctorat d'Auguste Laurent *et la* théorie des combinaisons organiques (1836)' *Bulletin de la Société chimique,* supplement May 1954: D-31–D-39.

Jacques, Jean (1987) *Berthelot. Autopsie d'un mythe.* Paris, Belin.

Jacques, Jean (1991) 'Professeurs et marchands' *Culture technique,* No. 23 'La chimie, ses industries et ses hommes': 46–52.

Jacques, Jean (1981) *Les confessions d'un chimiste ordinaire,* Paris, Seuil.

Jensen, Pablo (2001) *Entrer en matière. Les atomes expliquent-ils le monde?,* Paris, Seuil.

Joly, Bernard (1996) 'L'édition des Cours de chymie aux XVII⁰ et XVIII⁰ siècles: Obscurités et lumières d'une nouvelle discipline scientifique' *Archives et bibliothèques de Belgique,* **51**: 57–81.

Joly, Bernard (2000) 'Descartes *et la* chimie' in *L'esprit cartésien,* Bernard Bourgeois and Jacques Havet (eds.), Paris, Vrin, 2000, Vol. 1, pp. 216–221.

Joly, Bernard (2001) 'La théorie des cinq éléments d'Etienne de Clave' *Corpus, revue de philosophie,* **39**: 9–44.

Jones, Richard, L. (2004) *Soft Machines,* Oxford, Oxford University Press.

Joy, Bill (2000) 'Why the Future doesn't need us' *Wired*, **8** (www.wired.cpm/wired/archive/8.04/joy.html).

Kant, Immanuel (1963) *Critique of Pure Reason*, transl. Norman Kemp Smith, London, Macmillan, first published in 1781.

Kekulé, August (1867) 'On some points of chemical philosophy' *The Laboratory*, I, July 27, 1867, reprinted in R. Anschütz, *August Kekulé*, Vol. 2, Berlin, 1929.

Kekulé, August (1861) *Lehrbuch der organischen Chemie, oder der Chemie der Kohlenstoffverbindungen*, 2 vols, Erlangen, Enke.

Kim, Mi Gyung (2000) 'Chemical analysis and the domains of reality: Wilhelm Homberg *Essais de chimie*, 1702–09' *Studies in History and Philosophy of Science* **31**: 37–69.

Kim, Mi Gyung (2003) *Affinity, that Elusive Dream. A Genealogy of the Chemical Revolution*, Cambridge, MA, The MIT Press.

Klein, Ursula (1994) 'Origin of the concept of chemical compound' *Science in Context*, **7**: 163–204.

Klein, Ursula (1996) 'The chemical workshop tradition and the experimental practice: discontinuities within continuities' *Science in Context,* **9**: 251–287.

Klein, Ursula (ed.) (2001) *Tools and Modes of Representation in the Laboratory Sciences*, Dordrecht, Kluwer Academic Publications.

Klein, Ursula, and Wolfgang Lefevre (2007) *Materials in Eighteenth-Century Science: A Historical Ontology,* Cambridge, MA, The MIT Press.

Knight, David M. (1967) *Atoms and Elements. A Study of Theories of Matter in England in the Nineteenth Century*, London, Hutchinson.

Knight, David M. (1978) *The Transcendental Part of Chemistry*, Dawson, Folkestone, Kent.

Kragh, Helge (1979) 'Niels Bohr's second atomic theory' *Historical Studies in the Physical Sciences*, **10**: 123–186.

Krimsky, Sheldon (2000) *Hormonal Chaos: The Scientific and Social Origins of the Environmental Endocrine Hypothesis*. Baltimore, Johns Hopkins University Press.

La Métherie, Jean-Claude (1786) 'Discours préliminaire contenant un précis des nouvelles découvertes' *Observations sur la physique*, **27**: 1–53.

Langlois, Charles-Victor and Charles Seignobos (1898) *Introduction aux études historiques*, Paris, Hachette.

Larrère, Raphael (2002) 'Agriculture, artificialisation ou manipulation de la nature?' *Cosmopolitiques*, **1**: 158–173.

Laszlo, Pierre (2000) *Miroir de la chimie*, Paris, Seuil.

Laszlo, Pierre (2001) 'Handling proliferation,' *Hyle*, **7**, No. 2: 125–140.

Laszlo, Pierre (2004) *Le phénix et la salamandre. Histoire de sciences*, Paris, Le Pommier.

Latour, Bruno (1979) *Laboratory Life*, Los Angeles, Sage.

Latour, Bruno (1987) *Science in Action*, Milton Keynes, Open University Press.

Latour, Bruno (1996) *Petite réflexion sur le culte moderne des dieux faitiches*, Paris, Synthélabo.

Latour, Bruno (2001) *L'espoir de Pandore. Pour une version réaliste de l'activité scientifique*, Paris, La découverte.

Laurent, Auguste (1837) *Recherches diverses de chimie organique*, PhD in chemistry and physics, Paris Faculty of Sciences, 20 December 1837.

Laurent, Auguste (1854) *Méthode de chimie*, Paris, Mallet-Bachelet.

Lavoisier, Antoine-Laurent (1789) *Traité élémentaire de chimie*, Paris, Cuchet. Transl. Robert Kerr, *The Elements of Chemistry*, London, Kettilby, 1790.

Lavoisier, Antoine-Laurent (1862–1896) *Œuvres*, 6 vols, Paris, Imprimerie impériale and Imprimerie nationale.

Le Chatelier, Henry (1925) *Science et Industrie*, Paris, Flammarion.

Lecourt, Dominique (1996) *Prométhée, Faust, Frankenstein. Fondements imaginaires de l'éthique*, Paris, Synthélabo.

Legay, Natalie (1998) 'Chimie industrielle et principe de précaution' in Gérard Mondello (ed.) *Principe de précaution et industrie*, Paris, L'harmattan, pp. 120–151.

Lehn, Jean-Marie (1985) 'Supramolecular Chemistry, Receptors, Catalysts and Carriers' *Science*, **227,** 22 February: 849–856.

Lehn, Jean-Marie (1995) *Supramolecular Chemistry*, Weinheim, VCH.

Lehn, Jean-Marie (2003) 'Une chimie supramoléculaire foisonnante' *La lettre de l'Académie des sciences*, No. 10: 12–13.

Lequan, Mai (2000) *La chimie selon Kant*, Paris, PUF.

Lemery, Nicolas (1675) *Cours de chymie contenant la manière de faire les opérations qui sont en usage dans la médecine* Paris, Jacques Langlois fils.

Lemery, Nicolas (1677) *A Course of Chymistry*, transl. Walter Harris, London, Kettilby.

Levere, Trevor (1971) *Affinity and Matter. Elements of Chemical Philosophy 1800–1865*, Oxford, Clarendon Press.

Levi, Primo (1975) *Il Sistema periodico*, Torino, Einaudi. Transl. *The Periodic Table*, London, Abacus, 1985.

Levi, Primo (1978) *La chiave a stella*, Torino, Einaudi. Transl. *The Monkey's Wrench*, London, Penguin Books, 1995.

Lévy, Monique (1979) 'Les relations entre chimie et physique et le problème de la réduction' *Epistemologia*, **2**: 337–370.

Locke, John (1689) *An Essay Concerning Human Understanding*, reprint Oxford, Clarendon Press, 1975.

Lovelock, James (1979) *Gaia: A new look at life on earth*, Oxford, Oxford University Press.

Lucretius (50 BCE) *De Rerum Natura*, available in English translation on http://classics.mit.edu.

Lüthy, Christopher, John E. Murdoch, W. R. Newman (eds.) (2001) *Late Medieval and Early Modern Corpuscular Matter Theories*, Leiden, Brill.

Malabou, Catherine (2000) *Plasticité*, Paris, édition Léo Scheer.

Malley, Marjorie (1979) 'The Discovery of Atomic Transmutation: Scientific Style and Philosophy in France and Britain' *Isis*, **70**: 213–223.

Mann, Stephen, John Werbb and Robert J. P. Williams (eds.) (1989) *Biomineralization, Chemical and Biological Perspectives*, Weinheim, VCH.

Marinacci, Barbara (ed.) (1995) *Linus Pauling in his own Words: Selection from his Writings, Speeches and Interviews*, New York, Simon & Schuster.

Mauskopf, Seymour H. (ed.) (1993) *Chemical Sciences in the Modern World*, Philadelphia, University of Pennsylvania Press.

Mc Donough, William, and Michael Braungart (2002) *Cradle to Cradle. Remaking the Way we Make Things*, North Point Press.

Meikle, Jeffrey L. (1993) 'Beyond Plastics: Postmodernity and the Culture of Synthesis' Working paper No 5 in David E. Nye and Charlotte Granly (eds.), *Odense American Studies International Series*, Odense, Odense University, 1–15.

Meikle, Jeffrey L. (1995) *American Plastic. A Cultural History*, New Brunswick, Rutgers University Press.

Meikle, Jeffrey L. (1997) 'Material Doubts. The Consequences of Plastic' *Environmental History*, **2**, No. 3: 278–300.

Meinel, Christoph (1983) 'Theory or practice? The eighteenth-century debate on the scientific status of chemistry' *Ambix*, **3**: 121–132.

Mendeleev, Dmitrii (1871) 'La loi périodique des éléments chimiques' tr. fr. *Le Moniteur scientifique*, 1879, **21**: 693–745.

Mendeleev, Dmitrii (1889) 'The periodic law of the chemical elements' (Faraday Lecture) *Journal of the Chemical Society*, **55**: 634–656.

Merton, Robert K. (1942) 'The ethos of science' in P. Sztompka (ed.) *On the Social Structure of Science*, Chicago, The University of Chicago Press, 1972, pp. 267–276.

Metzger, Hélène (1923) *Les doctrines chimiques en France du début du XVIIe à la fin du XVIIIe siècle*, reprint Paris, Blanchard, 1969.

Metzger, Hélène (1930) *Newton, Stahl, Boerhaave et la doctrine chimique*, Paris, Félix Alcan.

Metzger, Hélène (1930) 'La chimie' in M. E. Cavaignac (ed.) *Histoire du monde*, T XIII : *La civilisation européenne moderne*, Paris, E. de Boccard, pp. 1–169.

Metzger , Hélène (1935) *La philosophie de la matière chez Lavoisier*, Paris, Hermann.

Meyer-Thurow, G. (1982) 'The industrialization of invention: A case study from the German chemical industry' *Isis*, **73**: 361–81.

Meyerson, Emile (1921) *De l'explication dans les sciences*. Paris, Payot. Transl. *Explanation in the Sciences*, Dordrecht, Kluwer, 1991.

Morrell, Jack (1972) 'The chemist breeders: The research schools of Liebig and Thomas Thomson' *Ambix*, **19**: 3–46.

Morris, Peter (ed.) (2002) *From Classical to Modern Chemistry. The Instrumental Revolution*, Cambridge, Royal Society of Chemistry, 2002.

Mosini, Valeria (ed.) (1996) *Philosophers in the Laboratory,* Roma, Euroroma.

Mossman, Susan and Peter Morris (eds.) (1994) *The Development of Plastics*, London, The Science Museum.

Multhauf, Robert P (1966) *The Origins of Chemistry*, London, Oldbourne.

Nagel, Ernst (1961) *The Structure of Science*, New York, Harcourt.

Nagel, Ernst (1970) 'Issues in the logic of reductive explanations' in H. Kiefer, M. Munitz (eds.) *Mind, Science and History*, Albany, SUNY Press, pp. 117–137.

Needham, Paul (1996) 'Aristotelian Chemistry: A Prelude to Duhemian Metaphysics' *Studies in the History and Philosophy of Science*, **27**: 251–269.

Newman, William R. (1996) 'The alchemical sources of Robert Boyle's corpuscular philosophy' *Annals of Science* **53**: 567–585.

Newman, William R. (1989) 'Technology and the Alchemical Debate in the Late Middle-Ages' *Isis*, **80**: 423–445.

Newman, William R. (1991) *The Summa Perfectionnis of Pseudo-Geber,* Leiden, E. J. Brill.

Newman, William R. (2004) *Promethean Ambitions. Alchemy and the Art-Nature Debate*, Chicago, Chicago University Press.

Newman, William R. (2006) *Atoms and Alchemy, Chemistry and the Experimental Origins of the Scientific Revolution*, Chicago, University of Chicago Press.

Newman, William R., and Larry Principe (2003) *Alchemy tried in the Fire*, Chicago, University of Chicago Press.

Nieto-Galan, Agusti (2001) *Colouring Textiles. A History of Natural Dyestuffs in Industrial Europe*, Dordrecht, Boston, Kluwer Academic Publisher.

Newton Isaac (1730) *Opticks*, 4th edition (1st edition, 1704) reprint London, Dover Publication, 1979.

Nowotny, Helga, M. Gibbons and P. Scott (2001) *Re-thinking Science: Knowledge and the Public in an Age of Uncertainty,* Cambridge, Polity Press.

Nye, Mary-Jo (1983) *The Question of the Atom: From the Karlsruhe Congress to the Solvay Conference, 1890–1911*, Los Angeles, Tomash.

Nye, Mary-Jo (1993) *From Chemical Philosophy to Theoretical Chemistry: Dynamics of Matter and Dynamics of Discipline, 1800–1950*, Berkeley, University of California Press.

Observatoire français des techniques avancées (2001) *Biomimétisme et matériaux*, Paris, éditions TEC-DOC.

Ostwald, W. (1895) 'Ueber die Überwindung des wissenschaftlichen Materialismus, *Verhanlungen der Gessellschaft deutscher Naturforscher und Ärtze*, 155–168, transl. (French) 'La déroute de l'atomisme contemporain' *Revue générale des sciences pures et appliquées*, **21** (15 novembre 1895): 935–948.

Paneth, Friedrich A. (1931) 'The epistemological status of the concept of element' *British Journal for the Philosophy of Science*, **13** (1962): 1–14, 144–160, reprinted in *Foundations of Chemistry*, **5** (2003): 113–145 (original German lecture given in 1931).

Pauling, Linus (1950) *College Chemistry*, New York, Freeman.

Perrin, Jean (1913) *Les atomes*, reprint Paris, Gallimard, 1970.

Pestre, Dominique (2003) *Science, argent et politique. Un essai d'interprétation*, Paris, INRA.

Petroski, Henry (1992) *To Engineer is Human. The Role of Failure in Successful Design*. New York. Vintage Books.

Polanyi, Michael (1958) *Personal Knowledge. Towards a Post-Critical Philosophy*, London, Routledge and Kegan Paul.

Pouchard, Michel (2003) 'Chimie omniprésente, sa force, ses faiblesses' *La lettre de l'Académie des sciences*, **10**: 4–7.

Powers, Richard (1998) *Gain*, New York, Picador.

Principe, Lawrence (1998) *The Aspiring Adept. Robert Boyle and His Alchemical Quest*, Princeton NJ, Princeton University Press.

Psarros, Nikos (1998) 'What has Philosophy to Offer to Chemistry' *Foundations of Science* **1**: 183–202.

Quine William O. (1961) *From a Logical Point of View: 9 Logico-Philosophical Essays*, 2nd edition, Cambridge MA, Harvard University Press.

Ramberg, Peter (2000) 'The Death of Vitalism and the Birth of Organic Chemistry: Wölher's Urea Synthesis and the Disciplinary Identity of Chemistry' *Ambix*, **47**: 170–195.

Ramberg, Peter (2001) 'Paper Tools and Fictional Worlds: Prediction, Synthesis, and Auxiliary Hypotheses in Chemistry' in U. Klein (ed.), *Tools and Modes of Representation in the Laboratory Sciences*, Dordrecht, Kluwer, pp. 61–78.

Ramsey, Jeffrey L. (1999) 'Recent Work in the History and Philosophy of Chemistry' *Perspectives on Science*, **6**: 409–426.

Rees, Martin (2003) *Our Final Hour. A Scientist's Warning, How Terror, Error, and Environmental Disaster Threaten Humankind's Future In This Century — On Earth and Beyond*, New York, Basic Books.

Reinhardt, Carsten (1998) 'An instrument of corporate strategy: The Central Research Laboratory at BASF 1868–1890' in Ernst Homburg, A. Travis and H. Schröter (eds.), *The Chemical Industry in Europe, 1850–1914: Industrial Growth, Pollution, and Professionalization*, Dordrecht, Kluwer, pp. 239–260.

Reinhardt, Carsten (2006) *Shifting and Reappraising. Physical Methods and the Transformation of Modern Chemistry*, Dagamore Beach, Science History Publications.

Renault, Emmanuel (2003) *Philosophie chimique. Hegel et la science dynamiste de son temps*, Bordeaux, Presses universitaires de Bordeaux.

Rey, Abel (1908) *L'énergétique et le mécanisme au point de vue des conditions de la connaissance*, Paris, Félix Alcan.

Rhees, David J. (1993) 'Corporate Advertising, Public Relations and Popular Exhibits: The Case of Du Pont' in B. Schroeder-Gudehus (ed.), *Industrial Society and its Museums 1890–1990,* London, Harwood Academic Publishers, pp. 67–76.

Rip, Arie (1991) 'The danger culture of industrial society' in R. E. Kasperson and P. J. M. Stallen (eds.), *Communicating Risks to the Public*, Dordrecht, Kluwer, pp. 345–365.

Riskin, Jessica (2002) *Science in the Age of Sensibility. The Sentimental Empiricists of the French Enlightenment*, Chicago, University of Chicago Press.

Roberts, Lissa (1991) 'Setting the Table: the Disciplinary Development of Eighteenth-Century Chemistry as read through the Changing Structure if its Tables' in Peter Dear (ed.), *The Literary Structure of Scientific Argument*, Philadelphia, University of Pennsylvania Press, pp. 99–132.

Roberts, Lissa (1995) 'The death of the sensuous chemist: The new chemistry and the transformation of sensuous technology' *Studies in the History and Philosophy of Science*, **26**, No. 4: 503–529.

Rocke, Alan J. (1984) *Chemical Atomism in the Nineteenth Century. From Dalton to Cannizzaro*, Columbus, Ohio State University Press.

Rocke, Alan J. (1993) *The Quiet Revolution: Hermann Kolbe and the Science of Organic Chemistry*, Berkeley, University of California Press.

Rocke, Alan J. (2001) *Nationalizing Science. Adolphe Wurtz and the Battle for French Chemistry*, Cambridge MA, MIT Press.

Rousseau, Jean-Jacques (n.d.) *Institutions chymiques*, reprint Paris, Fayard, 1999.

Roco, M. C., R. S. Williams, and P. Alivisastos (2000) *Nanotechnology. Research Directions IWGN Interagency Working Group on Nanoscience Workshop Report*, Dordrecht, Boston, Kluwer.

Rouxel, J., M. Tournoux, and R. Brec (eds.) (1994) *Soft Chemistry Routes to New Materials,* Switzerland, Trans. Tech. Publications.

Russell, Colin (1987) 'The changing role of synthesis in organic chemistry' *Ambix*, **34**: 169–180.

Ruthenberg, Klaus (1993) 'Friedrich Adolph Paneth (1887–1938)' *Hyle, An International Journal in the philosophy of chemistry*, **3**: 103–106.

Sainte-Claire Deville, Henri (1886–1887) 'Leçons sur l'affinité chimique' in *Leçons de la Société chimique de Paris*, Paris.

Scheidecker, Myriam (1997) 'A. E. Baudrimont (1806–1880): Les liens entre sa chimie et sa philosophie des sciences' *Archives internationales d'histoire des sciences*, **47**: 26–56.

Schummer, J. (2003) 'The Philosophy of Chemistry' *Endeavour*, **27**: 37–41.

Schummer, J. (2003b) 'The Notion of Nature in Chemistry' *Studies in History and Philosophy of Science*, **34**: 705–736.

Schummer, J., B. Bensaude-Vincent, and B. Van Tiggelen (eds.) (2007) *The Public Image of Chemistry*, Singapore, World Scientific Publishing.

Scerri, Eric R. (2000) 'The Failure of Reduction and How to Resist Disunity of the Sciences in the Context of Chemical Education' *Science & Education*, **9**: 405–425.

Scerri, Eric R., and Lee McIntyre (1997) 'The Case For the Philosophy of Chemistry' *Synthese*, **111**: 213–232.

Scerri, Eric R. (2007) *The Periodic Table. Its Story and Its Significance*, Oxford, Oxford University Press.

Serres, Michel (1990) *Le contrat naturel*, Paris, François Bourin.

Shapin, Steven and Simon Schaffer (1985) *Leviathan and the Air Pump: Hobbes, Boyle, and the Experimental Life*, Princeton, Princeton University Press.

Siegfried, Robert (2002) *From Elements to Atoms, A History of Chemical Composition*, Philadelphia, American Chemical Philosophy.

Siegfried, Robert, and Dobbs, Betty (1968) 'Composition: A Neglected Aspect of the Chemical Revolution' *Annals of Science*, **24**: 275–293.

Simões, Ana (2002) 'Dirac's claim and the chemists' *Physics in Perspective*, **4**: 253–266.

Simon, Jonathan (2002) 'Analysis and the Hierarchy of Nature in Eighteenth-Century Chemistry' *British Journal for the History of Science*, **35**: 1–16.

Simon, Jonathan (2002b) 'Authority and Authorship in the Method of Chemical Nomenclature' *Ambix*, **49**: 207–227.

Simon, Jonathan (2005) *Chemistry, Pharmacy and Revolution in France, 1777–1809*, Aldershot, Ashgate, 2005.

Smalley, Richard (2001) 'Of Chemistry, Love and Nanorobots' *Scientific American*, September: 76.

Smith, Pamela H. (1994) *The Business of Alchemy: Science and Culture in the Holy Roman Empire*, Princeton, Princeton University Press.

Stahl, Georg Ernst (1730) *Philosophical Principles of Universal Chemistry*, London, Osborn and Longman. A translation by Peter Shaw of *Fundamenta Chymiae Dogmaticorationalis et Experimentalis*, Nuremberg, 1723.

Stengers, Isabelle (1995) 'Ambiguous Affinity: The Newtonian Dream of Chemistry in the Eighteenth-Century' in Michel Serres (ed.) *A History of Scientific Thought*, Oxford, Blackwell, pp. 372–400.

Stengers, Isabelle (1997) *Cosmopolitiques*, Paris, éditions Synthélabo, 7 vols.

Stengers, Isabelle and B. Bensaude-Vincent (2003) *Cent mots pour commencer à penser les sciences*, Paris, Seuil.

Thenard, Jacques-Louis (1834–1836) *Traité élémentaire de chimie théorique et pratique*, 6th edition, Paris, Gauthier-Villars.

Travis, Anthony S., Willem J. Hornix, and Robert Bud (eds.) (1992) *Organic Chemistry and High Technology, 1850–1950,* Special Issue of *The British Journal for the History of Science*, **25**, March 1992.

Urbain, Georges (1921) *Les disciplines d'une science,* Paris, Doin.

Urbain, Georges (1925) *Les notions fondamentales d'élément chimique et d'atome*, Paris, Gauthier-Villars.

Van Brakel, Jap (2000) *Philosophy of Chemistry*, Leuven, Leuven University Press.

Van Spronsen, J. W. (1969) *The Periodic System. A History of the First Hundred Years*, Amsterdam, Elsevier.

Whitesides, George, J. P. Mathias, and C. T. Seto (1991) 'Molecular self-assembly and nanochemistry: A chemical strategy for the synthesis of nanostructures' *Science*, **254**: 1312–1319.

Whitesides, George M. (1995) 'Self-Assembling Materials' *Scientific American*, September: 146–149.

Whitesides, George M. (2001) 'The Once and Future Nanomachine' *Scientific American*, September: 78–83.

Woodward, Robert Burns (1956) 'Synthesis' in Alexander Todd (ed.) *Perspectives in Organic Chemistry*, New York, Interscience Publishers, pp. 155–184.

Wurtz, C. Adolphe (1868–78) *Dictionnaire de chimie pure et appliquée*, Paris, Hachette, 3 vols.

Wurtz, C. Adolphe (1877) 'Réponse à M. Berthelot sur l'atome' *Comptes-rendus de l'Académie des sciences*, **84**, No. 23: 1264–1268.

Xiang, X.-D. *et al.* (1995) 'A Combinatorial Approach to Materials Discovery' *Science*, **268**: 1738–1740.

Xiang, X.-D. (1999) 'Combinatorial Materials Synthesis and Screening: An Integrated Materials Chip Approach to Discovery and Optimization of Functional Materials' *Annual Review of Materials Science*, **29**: 149–171.

Zhang, Shuguang (2003) 'Fabrication of novel biomaterials through molecular self-assembly' *Nature Biotechnology*, **21**, No. 10: 1171–1178.

Website

History and philosophy of chemistry site: http://www.hyle.org/

INDEX